T0289729

MODELING AND ADVANCED TECHNIQUES IN MODERN ECONOMICS

MODELING AND ADVANCED TECHNIQUES IN MODERN ECONOMICS

Editors

Çağdaş Hakan Aladağ
Hacettepe University, Turkey

Nihan Potas
Ankara Hacı Bayram Veli University, Turkey

 **World Scientific**

NEW JERSEY · LONDON · SINGAPORE · BEIJING · SHANGHAI · HONG KONG · TAIPEI · CHENNAI · TOKYO

Published by

World Scientific Publishing Europe Ltd.

57 Shelton Street, Covent Garden, London WC2H 9HE

Head office: 5 Toh Tuck Link, Singapore 596224

USA office: 27 Warren Street, Suite 401-402, Hackensack, NJ 07601

Library of Congress Cataloging-in-Publication Data
Names: Aladağ, Çağdaş Hakan, editor. | Potas, Nihan, editor.
Title: Modeling and advanced techniques in modern economics / editors,
    Çağdaş Hakan Aladağ, Hacettepe University, Turkey,
    Nihan Potas, Ankara Hacı Bayram Veli University, Turkey.
Description: New Jersey : World Scientific, [2022] |
    Includes bibliographical references and index.
Identifiers: LCCN 2022000600 | ISBN 9781800611740 (hardcover) |
    ISBN 9781800611757 (ebook) | ISBN 9781800611764 (ebook other)
Subjects: LCSH: Econometric models. | Economics--Mathematical models.
Classification: LCC HB141 .M586 2022 | DDC 330.01/5195--dc23/eng/20220118
LC record available at https://lccn.loc.gov/2022000600

British Library Cataloguing-in-Publication Data
A catalogue record for this book is available from the British Library.

For any available supplementary material, please visit
https://www.worldscientific.com/worldscibooks/10.1142/Q0346#t=suppl

Desk Editor: Soundararajan Raghuraman

Typeset by Stallion Press
Email: enquiries@stallionpress.com

To my mother Rezzan Aladağ
and
my father Hikmet Feridun Aladağ

Çağdaş Hakan Aladağ

To my mother Prof. Dr. Şefika Şule Erçetin,
my sister Dr. Şuay Nilhan Açıkalın
and
my beloved children Mihri and Uraz for their eternal love

Nihan Potas

Preface

While the global economy stumbles under a pandemic crisis, understanding macroeconomic variables has become more vital not only for the governments and economy bureaucracies but also for firms and individuals. Thus, each of us needs to understand the impact of exogenous shocks on the global economy. Statistical thinking is one of the leading ways of improvement in this thinking. For statistical thinking, data that can be worked on and then, in order to make clear, understandable and explanatory results, statistical inference are required.

The way to shape and see the future is to collect data correctly, analyze it correctly and interpret the results correctly. There are two main elements at the base of this approach. These are Statistics and Economics. The aim of this book, briefly, is to introduce new approaches that can be used to shape and forecast the future by combining the two disciplines mentioned.

In this day and age, data is a vital asset for any organization, regardless of industry or size. The world is built upon data. It is not only data but also having the right kind of knowledge is critically important today. In order to obtain the right kind of knowledge from data, the process of analyzing data is very crucial for individuals, organizations or governments. This is why the concept of *Data Science* is so popular today. Nowadays, many researchers and practitioners from various fields are studying advanced data analysis techniques. One of these fields is Economics. For example, time series analysis is very important for economics. Time series, which is a collection of data points collected at constant time intervals, is one of the most important data types. People have tried to forecast time series by different methods for a long time. Forecasting is so essential since the prediction of future events is a critical input for many types of planning decision-making processes, with real-world applications in all areas. With the development of technology and algorithms, time series forecasting approaches are developing. Thus, it is possible to reach more accurate predictions of the future by using advanced forecasting approaches. In this manner, some chapters of this book are intended to be a valuable source of recent knowledge on advanced time series forecasting techniques. These chapters include applications of efficient, recent forecasting approaches. The readers can also find useful information on advanced time series forecasting techniques, such as artificial neural networks, deep learning, machine learning and chaotic time series. In addition to these time series applications, some chapters introduce some other recent data analysis methods such as fiducial method, a novel approach based on inverse Gaussian distribution and weighted superposition attraction–repulsion algorithm.

Trends in econometric analysis, especially in financial time series analysis, are moving toward artificial intelligence. Machine learning becomes handier when dealing with complex structures as their computational capacities increase. Moreover, in classical time series analysis, univariate or multivariate, linear time series methods may not capture these complex structures. Nowadays, deep neural networks are a popular tool in forecasting and classification. Neural networks are flexible and can be used in both regression and classification. They have good performance in predicting nonlinear and non-stationary time series. The learning algorithms used in financial time series or econometrics are not restricted to neural networks. In classification support vector machines, tree type algorithms and k-nearest neighbors are some of the most popular algorithms. In the regression part support vector regression, ridge regression, Elastic-Net regression and Lasso regression can be mentioned as some of the popular algorithms. In light of these, in some chapters of this book, the authors provide information on econometric and financial models as well as commonly used financial and economic variables and mention sources of relevant data. The readers can also find useful information about the financial and econometric models, stochastic financial models, machine learning and application of the models to financial and macroeconomic data.

Readers can also find useful information about the effect of the economy through theoretical, comparative and applied studies on agent-based models, nonlinear economic systems, evolutionary economics, econophysics, chaos theory, fractal analysis, neuroeconomics, fuzzy systems and network theory.

Economic agents, be they banks, firms, households, consumers or investors, act with strategic behavior and foresight by considering outcomes that might result from behavior they might undertake. Furthermore, they continually adjust their market operations, buying decisions, prices and forecasts to the situations that these operations, decisions, prices and forecasts together create. These topics add a layer of complexity to economics not found in the natural sciences.

The editors would also like to express their sincere thanks to all authors for their valuable contributions. We believe that this book is a very useful resource for the readers.

About the Editors

Çağdaş Hakan ALADAĞ is currently a Professor in the Faculty of Science, Department of Statistics, Operational Research Section at Hacettepe University. He has been a Visiting Scholar in Knowledge/Intelligence Systems Laboratory at University of Toronto, Toronto, Canada for one year. Aladağ has been teaching full-time at Hacettepe University, offering courses in operation research, computer programming, statistics, intelligent optimization techniques, soft computing methods and integer programming. His main fields of interest are time series forecasting, artificial neural networks, fuzzy time series, heuristic algorithms and advanced statistical approaches. He has given seminars on these topics at various universities in Turkey. Aladağ is an expert in heuristic algorithms and soft computing methods with special emphasis on time series forecasting. He has many publications including refereed journal papers, book chapters and books in these and related fields. He has also given many international conference presentations. Aladağ has contributed to the literature by proposing various advanced forecasting techniques, which are based on heuristic algorithms, computational methods and fuzzy logic, to solve different forecasting problems from different application areas. He has served as a member of the scientific committee of several international conferences, and he is a reviewer for many international journals and conferences. Moreover, he is a member of editorial boards of many esteemed international journals.

Nihan POTAS graduated from Başkent University, Department of Statistics and Computer Sciences in 2005. She has been working as an Assistant Professor at Ankara Hacı Bayram Veli University, Faculty of Economics and Administrative Sciences, Department of Healthcare Management. She had completed her master's education at Gazi University Department of Statistics, Faculty of Science. During her Ph.D., she enrolled in TUBITAK's international doctoral research fellowship program. Through her fellowship and doctoral research, she became a research scholar at the University of North Carolina at Chapel Hill in the Department of Biostatistics. She got her Ph.D. degree from Ankara University Department of Statistics, Faculty of Science. She has been the project manager, principal investigator and co-principal investigator of several national and international projects which are funded by grants. She is the founding board member of the International Science Association. She has many publications including refereed journal papers and book chapters. Her main interests are nonparametric statistics, categorical data analysis, survival analysis and statistical applications.

Contents

Chapter 1

Smart Growth Developments of European Union Members by Europe 2020 Strategy

Şahika Gökmen[*,†,‡,§] and Johan Lyhagen[†,¶]

[†]*Department of Statistics, Uppsala University, Sweden*
[‡]*Department of Econometrics,*
Ankara Hacı Bayram Veli University, Turkey

[§]*sahika.gokmen@statistics.uu.se*
[¶]*johan.lyhagen@statistics.uu.se*

The Europe 2020 Strategy is an agenda of 10-year aims for the European
Union (EU) countries in force for the current planning period (2010–
2020). This strategy includes five headline targets and three-leg growth
strategies, namely smart, sustainable and inclusive growth. In particular,
smart growth refers to developing an economy based on knowledge and
innovation. Hence, it can be considered as the priority target through
which the components of smart growth targets directly help the main aim
of becoming the world's most competitive and dynamic economy. The
motivation of this study is to examine both the annual and current situations of countries in the smart growth aspect. For this purpose, seven
variables and the 2010–2019 period were used for the analysis. Factor
analysis is the preferred research method because it can provide annual

---

[*]Corresponding author.

factor scores, which can be the basis of rankings of countries. According to the main results of the analysis for the most recent year (2019), Finland, Spain and France occupy the first three places, respectively, while Poland, Bulgaria and Luxembourg are at the bottom with Poland being the last.

1. Introduction

The financial crisis of 2008 is considered as the biggest since the Great Depression and is often named as the Great Recession because it spread as an economic collapse that severely affected many countries (Temin, 2010; Singhania and Anchalia, 2013; Eigner and Umlauft, 2015; Gopinath, 2020). The EU Commission, in its 2010 report, stated that the economic growth achieved in the last decade disappeared as the crisis turned into a global recession. The report also highlighted that GDP decreased by 4%, with 10% of the active population unemployed and that as much as 20 years of fiscal consolidation was wiped out in the EU. This situation is also considered as a threat and a setback factor for the Lisbon Strategy, by which the EU member states pursue to become the world's most competitive and dynamic economy. Accordingly, the EU Commission has learned the lessons from the crisis and has in the Europe 2020 Strategy set 10-year targets for achieving both the targets of the Lisbon Strategy and a sustainable future (EC, 2010). The three headlines identified in this strategy package are smart growth, sustainable growth and inclusive growth.

In general, the common objective of the three headlines is to create conditions for long-term sustainable growth in the EU members. In this context, five headline targets are defined in the Europe 2020 Strategy and are related to the following: (i) R&D; (ii) employment; (iii) education; (iv) green sustainability; (v) poverty and social exclusion. In particular, smart growth refers to developing an economy based on knowledge and innovation, sustainable growth refers to promoting a more resource-efficient, greener and competitive economy, and inclusive growth refers to fostering a high-employment economy delivering social and territorial cohesion (EC, 2010). According to this, it is recommended that the three-leg growth strategies should be supported by flagship initiatives from EU members and their local and regional authorities. The main targets and the flagship initiatives of the Europe 2020 Strategy are listed in Table A.1 in Appendix. As seen in Table A.1, smart growth strategies focus on the quality of education, research performance, innovation and knowledge transfer, supporting innovative ideas and a more competitive economy.

Based on the concepts of smart, sustainable and inclusive growth, it is seen that each of the headlines handles the growth differently, and so each of them affects the economy in different aspects. When considering smart growth, it is seen that it is the one with the largest share in GDP. The reason for this is that the developing economy definition of smart growth contains the concepts of knowledge, innovation, research and education aspects (Naldi *et al.*, 2015). These concepts also outline the flagship initiatives of smart growth which are called as innovation union, youth on the move and a digital agenda for Europe (EC, 2010) (see Table A.1 in Appendix). Additionally, if the headline targets are examined, it is seen that both the R&D and education targets are relevant to smart growth headlines directly, although all targets and initiatives are interrelated. According to these targets, bringing innovation to the economy and increasing R&D will increase both the competitiveness and new job possibilities of the EU, and improving education will decrease unemployment and poverty. The relationship between smart growth with GDP and workforce proves to be a stronger pillar than the other legs in the context of economic growth. Considering that the start of the Europe 2020 Strategy is the great financial crisis, strengthening GDP growth is one of the biggest steps toward achieving the EU's sustainable future and competitiveness goals. In this context, smart growth is the lead wheel of the headline targets in the Europe 2020 plan. The literature contains few studies focusing on one of the headline targets of the strategy, while most research works indicate that the smart growth pillar has a major impact on economic growth as well as employment and competitiveness (Cooke and de Propris, 2011; Kėdaitienė and Kėdaitis, 2012; Panitsides, 2014; Szymańska and Zalewska, 2018; Klikocka, 2019; Mazur-Wierzbicka, 2019; Cuestas *et al.*, 2021).

The leading role of smart growth is the main motivation of this chapter. The current situation of the EU countries in the progress toward the smart growth targets is quite informative for measuring the economic competitiveness. On the other hand, with the expiry of the Europe 2020 agenda in 2020, recent studies have been more informative about the situations of the member countries. Depending on the data publication process, the data period of the most recently published articles within this scope include at maximum the year 2018 (Landaluce-Calvo and Gozalo-Delgado, 2021; Kosareva and Krylovas, 2021). Therefore, by including both current and past smart growth pillar positions of the EU27 members (2010–2019), this study enriches the literature as it includes almost the end of the agenda period.

The layout of this chapter is determined based on the implementations of its purpose. Hence, in Section 2, the literature on the Europe 2020 Strategy is thoroughly reviewed. The empirical data, procedures of determining the variables and methodology are presented in Section 3, while the analysis findings and ranks of the EU27 countries based on the smart growth view are examined in Section 4. Finally, Section 5 includes the conclusions and the discussion.

2.   Literature Review

The Europe 2020 Strategy has received a lot of attention in the literature due to its focus on popular and important issues, inclusion of concrete goals over a certain period, and accessibility of data. Therefore, the literature has many different approaches to this field. First of all, if we consider the research in terms of the countries that are examined, while the majority present empirical research exhibiting advances made by member states in implementing the strategies (Naldi *et al.*, 2015; Szymańska and Zalewska, 2018; Klikocka, 2019; Copeland and Daly, 2012; Çolak and Ege, 2013; Balcerzak, 2015; Stec and Grzebyk, 2016; Fura *et al.*, 2021; Širá *et al.*, 2021), some of the research works are focused on the current situations of selected country or countries (Panitsides, 2014; Mazur-Wierzbicka, 2019; Bonsinetto and Falco, 2013; Kurushina and Kurushina, 2014; Liobikienė *et al.*, 2016; Radulescu *et al.*, 2018; Naterer *et al.*, 2018; Furmankiewicz *et al.*, 2021). In one of these studies, Banelienė

(2013) preferred a different approach and compared the smaller EU countries with the rest, while Manafi and Marinescu (2013) compared Romania with EU countries. In the studies mentioned, various empirical methods, such as different modeling approaches, classification and ranking, are used. On the other hand, there are some papers (Baležentis *et al.*, 2011; Baležentis and Baležentis, 2011; Brauers *et al.*, 2012; Brauers and Zavadskas, 2013; Ture *et al.*, 2019; Fedajev *et al.*, 2020) that used optimization methods to examine the country rankings regarding the Europe 2020 Strategy.

Additionally, most of the papers (Klikocka, 2019; Balcerzak, 2015; Rogge and Konttinen, 2018; Radulescu *et al.*, 2018; Leon and Nica, 2011; Rappai, 2016) prefer a dynamic approach to analyze the implementation of the targets over a specific time period, while others (Fura *et al.*, 2017; Manafi and Marinescu, 2013; Roman *et al.*, 2013) examine the targets at a specific time point. Also, researchers focused on measuring the competitiveness of EU countries via certain defined indexes. For instance, Pasimeni (2013) defined indexes for each main pillar, which he named "smart growth index, sustainable growth index and inclusive growth index" and then a "Europe 2020 index." Pasimeni (2013; 2012) quantified, measured and monitored the current situation toward the targets of the Europe 2020 Strategy of EU members. Pasimeni and Pasimeni (2016) estimated a multiple linear regression with the Europe 2020 index as the dependent variable and the economic and institutional factors as explanatory. They analyzed the successes of the countries in this way, and based on the results, detection of important institutional variables was made. However, according to Rappai (2016), one of the excogitative things of the Europe 2020 index is that the index value can reach 1 even if a country does not achieve the goals of Europe 2020 since the measurements of the index depend on the performance of the best-performing country. Although Colak and Ege (2013) suggested an indicator that remedied this weakness, Rappai (2016) pointed out that the study by Colak and Ege (2013) does not take into account both the heterogeneity and correlations between the indicators and/or countries. In this way, Rappai (2016) recommends a measure based on Mahalanobis distance. In the following years, there have been studies that suggest and/or use indexes within this field (Molina *et al.*, 2015;

Becker *et al.*, 2020; Landaluce-Calvo and Gozalo-Delgado, 2021) while some authors prefer to use single indicators (Choi and Calero, 2013).

There are also studies which research the strategic and political points of the Europe 2020 Strategy. For instance, while Hobza and Mourre (2010) focus on the scenario-based analysis of strategies, Kedaitiene and Kedaitis (2012) concentrate on the macroeconomic effects of the strategy. Based on the paper of Kedaitiene and Kedaitis (2012), it is shown that the Europe 2020 indicators have a major impact on GDP growth of EU27 and specifically that the level of R&D and innovations are the strongest indicators. Additionally, while Břízová (2013) assessed whether the national targets are reasonable or not, Kedaitis and Kedaitiene (2014) argued whether the success of the strategy depends on the EU's internal and external policies. Apart from all these aforementioned papers, there are papers focusing on certain specific political sides and strategic points of view of the Europe 2020 Strategy (Naldi *et al.*, 2015; Naterer *et al.*, 2018; Baležentis and Baležentis, 2011; Makarovič *et al.*, 2014; Drumaux and Joyce, 2018; 2020).

Besides the vast number of papers mentioned above, there are also studies focusing on the three main pillars or specific initiatives of the Europe 2020 Strategy, such as the current study. From this perspective, it is recognized that most of the studies deal with the sustainable growth leg of Europe 2020 since the conceptions of sustainability, climate change and green development are some of the most striking issues nowadays (Širá *et al.*, 2021; Bonsinetto and Falco, 2013; Liobikienė *et al.*, 2016; Naterer *et al.*, 2018; Thollander *et al.*, 2013; Liobikienė and Butkus, 2017; Moreno and García-Álvarez, 2018). In addition, the inclusive growth is attractive for authors because it includes poverty and social exclusion. Examples include Böhme *et al.* (2011), Vanhercke (2011), Copeland and Daly (2012), Leschke *et al.* (2012), Manafi and Marinescu (2013), Dumitrescu (2016) and Rogge and Konttinen (2018). While Klikocka (2019) investigates simultaneously the inclusive and smart growth, Panitsides (2014) and Cuestas *et al.* (2021) concentrate specifically on the education targets of Europe 2020. Both papers highlight the importance of education for a knowledge-based economy. In addition to this, Klikocka (2019) emphasizes that the growth rate of R&D

is a significant priority for improving the EU economies. Similarly, Mazur-Wierzbicka (2019) paid attention to the significant correlation between economic growth in the EU and investmens in R&D. It is also mentioned that the intensity of R&D in the EU is still not comparable to other developed countries, such as the United States, China or Japan. Even though the EU is at a relatively high level in the global sector of R&D, this has been weakened since the increase of expenditure on R&D by China (in 2015, China devoted 2.07% of its GDP to R&D) (Mazur-Wierzbicka, 2019). On the other hand, Szymańska and Zalewska (2018) indicated that using a percentage of GDP is an uncertain headline indicator for investment in the R&D sector. According to political and strategical research, Naldi *et al.* (2015) analyzed the policies of smart growth and indicated that smart growth supports sustainable development. Besides, Markowska and Strahl (2012) examined regional smart growth and argued that regional smart growth covers the three pillars of innovation, creativity and smart specialization. Also, Cooke and Propris (2011) investigated the EU policies for promoting creative and cultural industries for Smart Europe. In total, it can be said that smart growth targets have a major impact on the GDP growth and hence are a way toward being more competitive in the international arena especially.

Among the large number of studies assessing the Europe 2020 Strategy, those carried out after the end of the target period of the strategy can certainly give more information on the achievement of the pillar targets. For this reason, it is important to understand whether achieving the targets will turn into factual progress of the European economies or just turn the entire program into an illusion (Moniz, 2011). However, not even the latest papers include the end of the period. For example, the important paper of Fedajev *et al.* (2020) is published in 2020 but does not include the relevant period. In their paper, which ranks the EU15 countries by using the Shannon entropy index, it is highlighted that a development gap still remains among the EU countries and that this gap comprises investments in R&D and development of renewable energy production. Also, based on their results, it is indicated that Sweden, Denmark and Austria are the best performers in the strategy implementation. Similarly, a dynamic composite indicator is created by Landaluce-Calvo and

Gozalo-Delgado (2021) and a EU2020 index is created by Becker *et al.* (2020). The findings of Becker *et al.* (2020), indicated that only Sweden and Denmark have met the targets (except the level of greenhouse gas emission). Landaluce-Calvo and Gozalo-Delgado (2021) found similar results. In addition, the data of Landaluce-Calvo and Gozalo-Delgado (2021) covered the period, 2009–2018, and for this reason, their research can be considered as one of the most up-to-date and comprehensive studies. Similarly, Kosareva and Krylovas (2021) examined the current situation of the EU member countries via a multiple-criteria decision-making approach, and they mentioned that Sweden, Finland, Denmark and Austria are the most successful ones. However, the data belong to the 2016–2018 period, and as the Europe 2020 agenda period is 2010–2020, this is a crucial limitation. There are also studies focusing on more specific areas; as examples, the following can be mentioned: background of the sociological discrimination against the elderly in the Europe 2020 Strategy (Przybysz *et al.*, 2021), educational performance (Cuestas *et al.*, 2021) and sustainable development (Širá *et al.*, 2021) in EU countries. Although there exists literature on the current situations of EU countries concerning specific subjects, no studies can be found examining smart growth for EU members. Therefore, this study can make an important contribution to the literature, both with the most up-to-date data (2010–2019 period) and especially by examining smart growth (which is one of the biggest leg for the GDP growth of the EU members).

3. Data and Method

3.1. *Data and variables*

This study uses yearly data over the period, 2010–2019, concerning the seven variables for the EU27 countries.[1] The variables and their relationship with the flagship initiatives can be found in Table 1. The Europe 2020 agenda and the literature review above are used to

---

[1]Greece is excluded from the dataset since all observations could not be obtained.

Table 1.   Indicators used in the study.

Flagship Initiatives	Variables	Headline Indicators
Economy & Finance	x_1	GDP growth rate
	x_2	Inflation
R&D	x_3	R&D expenditure (shared in GDP)
Demography	x_4	Proportion of the population aged 0–14 in the total population
	x_5	Proportion of the population aged 15–64 in the total population
Education	x_6	Early school leavers
	x_7	Tertiary education

select the variables, the main indicators of smart growth and represents economic, innovation, education and demographic status. The period is selected such that it covers the period of the Europe 2020 Strategy and uses the most recent data available. The data source is EUROSTAT.

3.2.   Method

As discussed in Section 3.1, the data consist of a panel of seven variables for 26 countries measured under 10 years, which sets the limits on the methods that can be used. We arrange the data in the wide format $x^T = \{x_{2011}^T, \ldots, x_{2019}^T\}$, where x_t is a vector of variables at time t. Note that x is observed 26 times, once for each country. The basis for our method is the factor analysis model:

$$x = \Lambda \xi + \delta$$

where x is a $p \times 1$ vector of indicators, ξ a $p \times 1$ vector of latent variables and δ a vector of disturbances. The parameter matrix Λ relates the observed indicator variables x to the latent ξ. We assume the standard assumptions that the vectors ξ and δ are independent multivariate normal with covariance matrices $E\xi\xi^T = \Phi$ and $E\delta\delta^T = \Psi$, respectively. Note that the independence assumption implies $E\delta\xi^T = 0$. From this, we can easily derive the implied covariance matrix of x as $\Sigma = \Lambda\Phi\Lambda^T + \Psi$. Estimation is done by maximizing the log likelihood:

$$\log L = -\frac{n}{2} \left[\log |\Sigma| + \mathrm{tr}(S\Sigma^{-1}) \right]$$

The panel, or repeated measurements, version of a factor analysis model imposes a special structure on Σ. If we assume that the process is time-invariant and uncorrelated through time, then Σ is a block diagonal and can be written as $\Sigma = I_T \otimes \Sigma_t$, where T is the number of time points and Σ_t is the implied covariance matrix for time t. The assumption that the covariance matrix is the same for all t motivates us to use the Kronecker form. Additionally, we assume a time-invariant correlation between the same variable measured at different time points while the other correlations are zero. For the estimation, we need the sample counterparts. Σ_t is estimated by the covariance matrix of the long-form data, i.e. each variable stacked in a long vector and denoted as S_t. Further, the covariance between a variable at different time points is estimated as the average of the $9 \times 8/2 = 36$ covariances. This is denoted as $\overline{\mathrm{Cov}(x_i)}$ for variable i. In total, the mathematical expression for the covariance matrix is

$$S = I_9 \otimes S_t + \sum_{i=1}^{7} \left(\mathbf{1}_9 \mathbf{1}_9^T - I_9 \right) \otimes \left(\boldsymbol{E}_i \boldsymbol{E}_i^T \right) \overline{\mathrm{Cov}(x_i)}$$

where $\mathbf{1}_9$ is a 9×1 vector of ones and $\boldsymbol{E}_i$ is a vector of zeroes except a one at position i. The advantage of our formulation is that S is positive definite and has the same structure as the theoretical counterpart Σ.

As we are interested in the latent factors, which are unobserved, we need to estimate them. This can be done in different ways; here, we use regression factor scores, introduced by Thomson (1934) and Thurstone (1935). With the matrices defined above, the factor scores are estimated by

$$\hat{\boldsymbol{\xi}} = \hat{\Phi}\hat{\Lambda}^T \hat{\Sigma}^{-1} \boldsymbol{x}$$

The 9×1 vector $\hat{\boldsymbol{\xi}}$ has elements corresponding to one country for the nine time points, so we can analyze the progress through time. As we have 26 observations of $\boldsymbol{x}$, one for each country, we can also make comparisons between the countries within each year.

Table 2.   Descriptive statistics of variables.

Variables	Max.	Countries	Min.	Countries	Mean	S.D.
x_1	25.20	Ireland$_{2015}$	−6.60	Cyprus$_{2013}$	2.39	2.69
x_2	6.10	Romania$_{2010}$ Ireland$_{2010}$	−1.60	Bulgaria$_{2014}$	1.53	1.38
x_3	3.71	Finland$_{2010}$	0.38	Romania$_{2014}$	1.61	0.89
x_4	21.50	Ireland$_{2012}$ Ireland$_{2014}$	13.20	Bulgaria$_{2010}$ Bulgaria$_{2011}$ Germany$_{2014}$ Germany$_{2015}$ Germany$_{2016}$ Italy$_{2019}$	15.68	1.69
x_5	6.00	Malta$_{2014}$	−4.00	Ireland$_{2010}$	1.15	1.47
x_6	28.30	Portugal$_{2010}$	2.80	Croatia$_{2014}$ Croatia$_{2015}$ Croatia$_{2016}$	10.20	4.99
x_7	58.80	Cyprus$_{2019}$	18.30	Romania$_{2010}$	39.23	9.89

4.   Results

First of all, the descriptive statistics of variables are obtained to provide preliminary information regarding the smart growth variables of the EU member states. The statistics based on the variables are provided by Table 2, and according to the table, the average GDP growth of the EU27 countries is obtained as 2.39 while the inflation rate is 1.53. The GDP growth fell to approximately −4.33 for the EU in 2009[2] after the crisis, and so, based on the updated value, the EU countries are making progress in being competitive economies. Here, Ireland reached the maximum growth rate in the agenda process and Cyprus the minimum in the first years of the agenda. It can be thought that the effects of the crisis persisted for a longer time on Cyprus. Similarly, Romania is seen as one of the most affected among the EU countries in the context of its inflation rate reaching the maximum value in 2010. Also, it is noted from Table 2 that

---

[2] *Source*: World Bank.

the EU countries have not reached the R&D target yet in general. The maximum value of this variable belongs to Finland in 2010 but not in recent years. As for the population target, which is about increasing the share of population aged 20–64 to at least 75%, it is preferred to examine this for both age groups of 0–14 and 15–64 in this study. Based on the descriptive statistics, the share of population aged 0–14 is much higher when compared with that aged 15–64. It can be interpreted that the population goal could not be accomplished at the end of the agenda, but it is possible to be attained in the long term. Finally, the education targets have been almost reached considering the average of the EU. Especially in the tertiary education target, the Cyprus' value is striking.

The situations and efforts of the countries are also important in the context of collaboration for accomplishing the targets of the Europe 2020 strategies. Hence, descriptive statistics for countries is given in Table A.2 in Appendix. Accordingly, the most remarkable results belong to x_7 (having completed tertiary education to at least 40%) as the average values of 15 countries are above the overall average. Likewise, 17 countries in the EU27 have achieved the early school leavers goal based on the 10-year averages. On the contrary, the population target variables have the worst values since only less than half of the EU27 countries are above the overall average. Also, only Finland and Sweden have higher values than the target of 3%, 3.1 and 3.25, respectively, while 11 countries are above the overall average (1.61). Furthermore, as mentioned above, the EU27 countries are growing economically and becoming more competitive. Most countries have remained below the overall average of GDP growth and inflation (14 and 16 countries, respectively) based on the 10-year averages.

Smart growth rankings of the EU27 countries are obtained based on the factor scores for each year, and they are displayed in Table 3. When the table is examined in general aspect, the most remarkable results belong to Portugal with three first places (in 2011, 2012 and 2015) during the Europe 2020 Strategy period. Considering that these years belong to the beginning of the agenda, it can be interpreted that the reason for this success is the lesser effect of the financial crisis compared to that on the rest of the countries in the smart growth context. The reason for this interpretation is that Portugal dropped to the 17th rank in 2019. Luxembourg faced a similar situation after it took first place two times (in 2016 and 2018). However, it is noteworthy that in 2015 and 2016, the country jumped incredibly from the last to the first rank and then fell to the third place from the end. Estonia, Slovakia and Netherlands have indicated similar leaps in different years too. It may also be noted that there is no such sudden decline in any of the countries.

According to Table 3, similar to the case of Portugal, Denmark and Lithuania fell to low rankings in recent years while both of them had frontier ranks in the first years of the agenda. This means that

Table 3. Annual smart growth rankings of the EU27 countries.

	2011	2012	2013	2014	2015	2016	2017	2018	2019
Austria	13	11	6	3	15	14	23	16	21
Belgium	**26**	21	20	6	17	3	10	7	12
Bulgaria	7	**26**	18	18	22	23	1	8	25
Croatia	18	14	4	**1**	14	25	19	15	18
Cyprus	2	2	**1**	2	10	**26**	24	11	23
Czechia	14	22	11	22	18	15	12	4	19
Denmark	5	9	2	15	4	7	21	**26**	15
Estonia	25	25	12	**26**	3	8	5	23	9
Finland	16	13	22	24	6	4	15	20	**1**
France	22	24	10	20	16	11	8	9	3
Germany	23	3	9	17	12	21	22	19	20
Hungary	11	6	13	13	20	24	16	17	14
Ireland	24	19	7	11	25	2	4	2	4
Italy	20	8	23	4	7	10	7	18	7
Latvia	4	10	8	16	24	17	9	21	10
Lithuania	8	7	14	7	2	12	25	22	22
Luxembourg	9	12	5	19	**26**	**1**	18	**1**	24
Malta	3	5	3	25	19	5	6	5	8
Netherlands	21	15	**26**	5	5	19	3	13	13
Poland	19	4	19	8	9	13	13	24	**26**
Portugal	**1**	**1**	17	9	**1**	16	14	14	17
Romania	6	18	15	10	21	22	20	10	6
Slovakia	10	20	21	12	11	6	**26**	3	11
Slovenia	17	16	16	21	13	9	2	25	16
Spain	15	17	24	14	8	18	11	12	2
Sweden	12	23	25	23	23	20	17	6	5

Note: The first and last positions are shown in bold.

these countries are making poor progress in terms of smart growth. Ireland had a reverse situation, and its rank reached fourth in 2019 although it had a low ranking in 2011. When talking about the progress of the countries, it also appears that the two countries, France and Sweden, are in a more consistent upward trend compared to the other EU27 countries, and they reached third and fifth places, respectively, in 2019.

In general, it is seen that Finland, Spain, France, Ireland and Sweden occupy the first five places, respectively. However, these countries, except Sweden, have the last positions in 2011. On the contrary, Cyprus was ranked 23rd position, though it was second in 2011. Also, Poland, Bulgaria, Luxembourg, Cyprus and Lithuania occupied the last five places from the worst, respectively, and Bulgaria, Luxembourg and Lithuania were ranked among the top 10 in the beginning.

5. Conclusion and Discussion

The main idea of this chapter is to investigate the progress of the EU27 countries upon their smart growth goals with factor analysis. For this purpose, seven variables and 10 years of data, which represent the smart growth goals of the Europe 2020 Strategy, are used for the analysis. Data are extracted for the 2010–2019 period from the EUROSTAT database. Through the factor analysis, yearly country factor scores are obtained and ranks are interpreted in this context.

Based on the main results, Finland, Spain and France had the first three places, while Poland, Bulgaria and Luxembourg had the last three places in smart growth. Besides, education targets are mostly achieved while population targets are not. However, the increase in the population ratio in the earlier age cohort is a sign that the population target can be achieved in the long term. In addition, relating this with previous works on ranking (Kosareva and Krylovas, 2021; Fura *et al.*, 2017; Kurushina and Kurushina, 2014; Fedajev *et al.*, 2020; Becker *et al.*, 2020), it is shown that, even though there are differences in the order of countries, the places of countries are almost the same. However, Sweden, Denmark and Austria are found to the best in accomplishing the Europe 2020 Strategy goals based on all these papers' findings. One of the reasons for these differences is that, as Kosareva and Krylovas (2021) mentioned, the countries such as Sweden, Denmark and Austria are paying close attention to the sustainable use of natural resources and solving education system problems. Even smart growth contains education targets, but it does not contain sustainability targets since it is under the sustainable growth leg of the Europe 2020 Strategy. Therefore, ranking for the three legs of the strategy can lead to higher places for countries in comparison with ranking for one leg of the strategy. In addition, none of these papers included the 2019 data in their dataset. In this study, Sweden is at the fifth place while Denmark is at 15th and Austria at 21st based on the 2019 findings. Also, considering the findings from the classification of unsuccessful countries in related studies (Fura *et al.*, 2017; Fedajev *et al.*, 2020; Grimaccia, 2021), Fura *et al.* (2017) and Fedajev *et al.* (2020) defined Bulgaria as the weakest, while Grimaccia (2021) defined Italy as the weakest in accomplishing the goals. Based on the current study, Bulgaria is at 25th position in 2019 while Italy is at seventh. Although the studies take into account different goals, the interpretation is that Bulgaria has not been able to successfully protect itself.

This study will contribute to the relevant literature as it includes the most updated period and considers the smart growth aspect of the Europe 2020 strategies (which is highly representative of power and competitiveness of an economy). Also, the findings provide precious knowledge for decision makers about their countries' situations in accomplishing the goals. Even though the final rankings of the EU27 countries could not be obtained since the data for 2020 is not available yet, the 2019 rankings are informative for the achievement of the goals. This is because the effects of the pandemic on the Europe 2020 Strategy are not known yet. Based on this, in the future, the impacts of the COVID-19 pandemic on the Europe 2020 Strategy should be investigated.

References

Balcerzak, A.P. (2015). Europe 2020 Strategy and structural diversity between old and new member states. Application of zero-unitarizatin method for dynamic analysis in the years 2004–2013. *Interdisciplinary Approach to Economics and Sociology*, **8**(2): 190–210.

Baležentis, A. and Baležentis, T. (2011). Framework of strategic management model for strategy Europe 2020. *Inzinerine Ekonomika-Engineering*, **22**(3): 271–282.

Baležentis, A., Baležentis, T. and Brauers, W.K. (2011). Implementation of the strategy Europe 2020 by the multi-objective evaluation method multimoora. *Ekonomie A Management*, **2**: 6–21.

Banelienė, R. (2013). Evaluation of the efficiency of economic policy under the Europe-2020 Strategy in small European Union countries. *Ekonomika*, **92**(2): 7–19.

Becker, W., Norlén, H., Dijkstra, L. and Athanasoglou, S. (2020). Wrapping up the Europe 2020 Strategy: A multidimensional indicator analysis. *Environmental and Sustainability Indicators*, **8**: 1–13.

Böhme, K., Doucet, P., Komornicki, T., Zaucha, J. and Swiatek, D. (September, 2011). How to strengthen the territorial dimension of Europe 2020 and the EU cohesion policy. Technical report, Ministry of Regional Development, Warsaw.

Bonsinetto, F. and Falco, E. (2013). Analysing Italian regional patterns in green economy and climate change. Can Italy leverage on Europe 2020 Strategy to face sustainable growth challenges? *Journal of Urban and Regional Analysis*, **5**(2): 123–142.

Brauers, W.K.M. and Zavadskas, E.K. (2013). Multi-objective decision making with a large number of objectives. An application for Europe 2020. *International Journal of Operations Research*, **10**(2): 67–79.

Brauers, W.K.M., Baležentis, A. and Baležentis, T. (2012). European union member states preparing for Europe 2020. An application of the multimoora method. *Technological and Economic Development of Economy*, **18**(4): 567–587.

Břízová, P. (2013). Europe 2020 Strategy: Are national goals reasonable? Master's thesis, Charles University.

Choi, A. and Calero, J. (2013). The contribution of the population of disabled people to the attainment of Europe 2020 Strategy headline targets. *Disability & Society*, **28**(6): 853–873.

Çolak, M.S. and Ege, A. (2013). An assessment of EU 2020 Strategy: Too far to reach? *Social Indicators Research*, **110**(2): 659–680.

Cooke, P. and De Propris, L. (2011). A policy agenda for EU smart growth: The role of creative and cultural industries. *Policy Studies*, **32**(4): 365–375.

Copeland, P. and Daly, M. (2012). Varieties of poverty reduction: Inserting the poverty and social exclusion target into Europe 2020. *Journal of European Social Policy*, **22**(3): 273–287.

Cuestas, J.C., Monfort, M. and Ordóñez, J. (2021). The education pillar of the Europe 2020 Strategy: A convergence analysis. *Empirica*, (forthcoming): 1–17.

Drumaux, A. and Joyce, P. (2018). The commission as part of the 'Centre of Government' for the Europe 2020 Strategy, In *Strategic Management for Public Governance in Europe*, pp. 203–232. Palgrave Macmillan UK, London.

Drumaux, A. and Joyce, P. (2020). New development: Implementing and evaluating government strategic plans the Europe 2020 Strategy. *Public Money & Management*, **40**(4): 294–298.

Dumitrescu, A.L. (2016). The social dimension of "Europe 2020" Strategy: The inclusive growth. *Knowledge Horizons. Economics*, **8**(1): 68–72.

EC, (2010). Europe 2020: A strategy for smart, sustainable and inclusive growth. *Working paper {COM (2010) 2020}*.

Eigner, P. and Umlauft, T.S. (2015). The great depression(s) of 1929–1933 and 2007–2009? Parallels, differences and policy lessons (July 1, 2015). *Hungarian Academy of Science MTA-ELTE Crisis History Working Paper No. 2*.

Fedajev, A., Stanujkic, D., Karabaevi, D., Brauers, W.K. and Zavadskas, E.K. (2020). Assessment of progress towards Europe 2020 Strategy targets by using the Multimoora method and the Shannon entropy index. *Journal of Cleaner Production*, **244**: 1–12.

Fura, B., Wojnar, J. and Kasprzyk, B. (2017). Ranking and classification of EU countries regarding their levels of implementation of the Europe 2020 Strategy. *Journal of Cleaner Production*, **165**: 968–979.

Furmankiewicz, M., Janc, K., Kaczmarek, I. and Solecka, I. (2021). Are rural stakeholder needs compliant with the targets of the Europe 2020 Strategy? Text mining analysis of local action group strategies from two polish regions. In *Hradec Economic Days*, Vol. 11, pp. 195–206, University of Hradec Kralove, Czech Republic.

Gopinath, G. (2020). The great lockdown: Worst economic downturn since the great depression, *IMF Blog*, **14**.

Grimaccia, E. (2021). Europe 2020 Strategy for a smart, inclusive and sustainable growth: A first evaluation. *Rivista Italiana di Economia Demografia e Statistica*, **75**(1): 65–76.

Hobza, A. and Mourre, G. (2010). Macroeconomic effects of Europe 2020: Stylised scenarios. *ECFIN Economic Brief*, **11**.

Kėdaitienė, A. and Kėdaitis, V. (2012). Macroeconomic effects of the Europe 2020 Strategy. *Socialiniai Tyrimai*, **4**(29): 5–19.

Kedaitis, V. and Kedaitiene, A. (2014). External dimension of the Europe 2020 Strategy. *Procedia-Social and Behavioral Sciences*, **110**: 700–709.

Klikocka, H. (2019). Assumptions and implementation of smart growth and inclusive growth targets under the Europe 2020 Strategy. *European Research Studies*, **22**(2): 199–217.

Kosareva, N. and Krylovas, A. (2021). Assessing the Europe 2020 Strategy implementation using interval entropy and cluster analysis for inter-relation between two groups of headline indicators. *Entropy*, **23**(345): 1–26.

Kurushina, E.V. and Kurushina, V.A. (2014). Evolution of economic development aims. Assessment of the smart growth. *Life Science Journal*, **11**(11): 517–521.

Landaluce-Calvo, M.I. and Gozalo-Delgado, M. (2021). Proposal for a dynamic composite indicator: Application in a comparative analysis of trends in the EU member states towards the Europe 2020 Strategy. *Social Indicators Research*, **154**(3): 1031–1053.

Leon, R. and Nica, P. (2011). Europe 2020 Strategy-forecasting the level of achieving its goals by the EU member states. *Management & Marketing*, **6**(1): 3–18.

Leschke, J., Theodoropoulou, S. and Watt, A. (2012). *A Triumph of Failed Ideas European Models of Capitalism in the Crisis*, Chapter: How do economic governance reforms and austerity measures affect inclusive growth as formulated in the Europe 2020 Strategy? pp. 243–252. ETUI, Brussels.

Liobikienė, G. and Butkus, M. (2017). The European Union possibilities to achieve targets of Europe 2020 and Paris agreement climate policy. *Renewable Energy*, **106**: 298–309.

Liobikienė, G., Butkus, M. and Bernatonienė, J. (2016). Drivers of greenhouse gas emissions in the baltic states: Decomposition analysis related to the implementation of Europe 2020 Strategy. *Renewable and Sustainable Energy Reviews*, **54**: 309–317.

Makarovič, M., Šušteršič, J. and Rončević, B. (2014). Is Europe 2020 set to fail? The cultural political economy of the EU grand strategies. *European Planning Studies*, **22**(3): 610–626.

Manafi, I. and Marinescu, D.E. (2013). The influence of investment in education on inclusive growth — empirical evidence from Romania vs. EU. *Procedia–Social and Behavioral Sciences*, **93**: 689–694.

Markowska, M. and Strahl, D. (2012). European regional space classification regarding smart growth level. *Comparative Economic Research-Central and Eastern Europe*, **15**(4): 233–247.

Markowska, M. and Strahl, D. (2019). Median classification of the European Union countries regarding the level of selected strategic goals implementation–dynamic approach. In *The 13th Professor Aleksander Zelias International Conference on Modelling and Forecasting of Socio-Economic Phenomena*, pp. 150–159, Poland.

Mazur-Wierzbicka, E. (2019). Smart growth as a challenge for Poland in the light of the Europe 2020 Strategy. *Journal of Economics & Management*, **37**: 87–106.

Molina, M.D.M.H., Fernández, J.A.S. and Martín, J.A.R. (2015). A synthetic indicator to measure the economic and social cohesion of the regions of Spain and Portugal. *rEviSta dE Economía mundial*, **39**: 223–239.

Moniz, A.B. (2011). *Brain Drain or Brain Gain*, Chapter: From the Lisbon Strategy to EU2020–Illusion or Progress for European Economies? pp. 53–80. Edition Sigma, Berlin.

Moreno, B. and García-Álvarez, M.T. (2018). Measuring the progress towards a resource-efficient European Union under the Europe 2020 Strategy. *Journal of Cleaner Production*, **170**: 991–1005.

Naldi, L., Nilsson, P., Westlund, H. and Wixe, S. (2015). What is smart rural development? *Journal of Rural Studies*, **40**: 90–101.

Naterer, A., Žižek, A. and Lavrič, M. (2018). The quality of integrated urban strategies in light of the Europe 2020 Strategy: The case of Slovenia. *Cities*, **72**: 369–378.

Panitsides, E.A. (2014). Europe 2020 — Practical implications for the Greek education and training system: A qualitative study. *Procedia-Social and Behavioral Sciences*, **140**: 307–311.

Pasimeni, P. (2012). Measuring Europe 2020: A new tool to assess the strategy. *International Journal of Innovation and Regional Development*, **4**(5): 365–385.

Pasimeni, P. (2013). The Europe 2020 index. *Social Indicators Research*, **110**(2): 613–635.

Pasimeni, F. and Pasimeni, P. (2016). An institutional analysis of the Europe 2020 Strategy. *Social Indicators Research*, **127**(3): 1021–1038.

Przybysz, K., Stanimir, A., Wasiak, M., *et al.* (2021). Subjective assessment of seniors on the phenomenon of discrimination: Analysis against the background of the Europe 2020 Strategy implementation. *European Research Studies Journal*, **24**(Special 1): 810–835.

Radulescu, M., Fedajev, A., Sinisi, C.I., Popescu, C. and Iacob, S.E. (2018). Europe 2020 implementation as driver of economic performance and competitiveness. Panel analysis of CEE countries. *Sustainability*, **10**(2): 566–586.

Rappai, G. (2016). Europe en route to 2020: A new way of evaluating the overall fulfillment of the Europe 2020 strategic goals. *Social Indicators Research*, **129**(1): 77–93.

Rogge, N. and Konttinen, E. (2018). Social inclusion in the EU since the enlargement: Progress or regress? *Social Indicators Research*, **135**(2): 563–584.

Roman, M.D., Manafi, I. and Marinescu, D. (2013). Is Europe 2020 a realistic strategy for sustainable growth after the crisis. In *Proceedings of the 9th International Conference on Applied and Theoretical Mechanics*, Dubrovnik.

Singhania, M. and Anchalia, J. (2013). Volatility in asian stock markets and global financial crisis. *Journal of Advances in Management Research*, **10**(3): 333–351.

Širá, E., Kotulič, R., Kravčáková Vozárová, I. and Daňová, M. (2021). Sustainable development in EU countries in the framework of the Europe 2020 Strategy. *Processes*, **9**(3): 443–460.

Stec, M. and Grzebyk, M. (2016). The implementation of the strategy Europe 2020 objectives in European union countries: The concept analysis and statistical evaluation. *Quality & Quantity*, **52**(1): 119–133.

Szymańska, A. and Zalewska, E. (2018). Towards the goals of the Europe 2020 Strategy: Convergence or divergence of the European union countries? *Comparative Economic Research*, **21**(1): 67–82.

Temin, P. (2010). The great recession & the great depression. *Daedalus*, **139**(4): 115–124.

Thollander, P., Rohdin, P., Moshfegh, B., Karlsson, M., Söderström, M. and Trygg, L. (2013). Energy in swedish industry 2020–current status, policy instruments, and policy implications. *Journal of Cleaner Production*, **51**: 109–117.

Thomson, G. (1934). The meaning of i in the estimate of g. *British Journal of Psychology*, **25**(1): 92–99.

Thurstone, L. (1935). The vectors of mind: Multiple-factor analysis for the isolation of primary traits. University of Chicago Press, USA.

Ture, H., Dogan, S. and Kocak, D. (2019). Assessing Euro 2020 strategy using multi-criteria decision making methods: Vikor and Topsis. *Social Indicators Research*, **142**(2): 645–665.

Vanhercke, B. (2011). *Social Developments in the European Union*, Chapter: Is the social dimension of Europe 2020 an oxymoron? ETUI, Brussels.

Appendix

Table A.1.   Pillar targets and flagship initiatives of the Europe 2020 Strategy.

Headline Targets

i. Raising the employment rate of the active population (aged 20–64) to at least 75%.
ii. Achieving the proportion of the R&D investments in GDP to 3%.
iii. Meeting the "20/20/20" climate/energy targets.
iv. Reducing the proportion of early school leavers to 10% and increasing the proportion of the population aged 30–34 having completed tertiary education to at least 40%.
v. Decreasing the number of Europeans who are at risk of poverty or social exclusion by at least 20 million people.

Smart Growth
• Innovation Union improving research and development activities and innovation to solve problems.
• Youth on the move improving education quality and enhancing the performance of education systems and reaching higher education level.
• A digital agenda for Europe — speeding up the roll out of high-speed internet and reaping the benefits of a digital single market for household firms.

Sustainable Growth
• Resource-efficient Europe helping to decouple economic growth from the use of sources by decarbonizing the economy, increasing the use of renewable resources, modernizing the transport sector and promoting energy efficiency.
• An industrial policy for the globalization era improving the business environment and supporting the development of a strong and sustainable industrial base able to compete globally.

Inclusive Growth
• An agenda for new skills and jobs modernizing labor markets for increasing labor participation and better matching labor supply and demand.
• European platform against poverty ensuring social and territorial cohesion are widely shared and people experiencing poverty and social exclusion are enabled to live in dignity and take an active part in society.

Source: Europe 2020 Strategy, 2010 (EC, 2010).

Table A.2. Descriptive statistics of smart growth for the EU27 countries.

Countries	x_1 Mean	S.D.	x_2 Mean	S.D.	x_3 Mean	S.D.	x_4 Mean	S.D.	x_5 Mean	S.D.	x_6 Mean	S.D.	x_7 Mean	S.D.
Austria	1.55	0.94	1.91*	0.81	2.99*	0.17	14.47	0.20	0.86	0.24	7.58	0.52	34.29	8.07
Belgium	1.59	0.67	1.81*	0.93	2.44*	0.24	16.98*	0.04	0.45	0.58	10.18	1.52	44.15*	1.90
Bulgaria	2.37	1.48	1.15	1.92	0.72	0.14	13.77	0.42	1.47*	1.82	12.88*	0.65	30.74	2.65
Croatia	1.04	2.18	1.2	1.24	0.85	0.10	14.82	0.35	0.21	1.69	3.76	1.05	28.52	4.03
Cyprus	1.37	4.20	0.86	1.72	0.51	0.10	16.46*	0.32	0.44	0.91	8.96	2.30	52.14*	4.67
Czechia	2.43*	1.94	1.66*	1.06	1.77*	0.20	15.11	0.54	1.95*	0.87	5.86	0.71	29.05	4.96
Denmark	1.92	0.94	1.09	0.98	2.96*	0.05	17.19*	0.57	−0.29	1.42	9.24	1.29	45.15*	3.06
Estonia	3.74*	1.84	2.6*	1.67	1.61*	0.35	15.83*	0.44	1.54*	1.02	1.86	0.82	43.81*	3.16
Finland	1.21	1.71	1.49	1.14	3.1*	0.39	16.35*	0.17	0.63	1.15	8.87	0.93	45.56*	0.86
France	1.38	0.67	1.28	0.79	2.21*	0.03	18.45*	0.22	0.63	0.26	9.9	1.69	44.42*	1.45
Germany	1.93	1.36	1.43	0.68	2.94*	0.14	13.39	0.17	0.82	0.40	10.43*	0.73	32.63	1.85
Hungary	2.79*	2.05	2.52*	1.97	1.31	0.14	14.51	0.10	3.17*	1.64	11.81*	0.55	31.71	2.77
Ireland	6.27*	7.54	0.4	0.93	1.33	0.26	21.19*	0.33	−0.45	1.96	7.62	2.60	53.74*	1.72
Italy	0.26	1.53	1.23	1.14	1.33	0.10	13.77	0.30	0.91	0.89	15.6*	1.85	24.24	2.95
Latvia	2.55*	2.87	1.45	1.71	0.61	0.10	14.93	0.67	1.17	1.12	9.89	1.49	40.26*	3.99
Lithuania	3.56*	1.26	1.83*	1.57	0.93	0.10	14.82	0.17	2.53*	0.40	5.83	1.23	53.24*	5.60
Luxembourg	3.11*	1.62	1.76*	1.22	1.3	0.10	16.78*	0.58	1.09	1.02	6.92	1.14	52.11*	3.30
Malta	5.57*	2.52	1.61*	0.77	0.65	0.10	14.42	0.50	4.55*	1.44	20.16*	2.27	29.65	5.02
Netherlands	1.44	1.17	1.5	1.10	2.06*	0.17	16.8*	0.60	0.93	0.39	8.43	0.97	45.35*	3.47
Poland	3.63*	1.46	1.51	1.58	0.97	0.17	15.15	0.14	1.86*	0.20	5.32	0.28	41.9*	4.08
Portugal	0.84	2.37	1.22	1.19	1.36	0.10	14.47	0.56	0.41	0.94	17.08*	5.65	30.95	3.83
Romania	3.1*	2.90	2.75*	2.44	0.47	0.04	15.66	0.14	1.47*	0.78	17.8*	1.21	23.61	2.71
Slovakia	2.99*	1.50	1.58*	1.66	0.83	0.14	15.44	0.14	1.17	0.66	6.87	1.56	29.48	6.23
Slovenia	1.92	2.33	1.35	1.15	2.02*	0.24	14.62	0.36	0.95	1.94	4.49	0.37	41.46*	3.55
Spain	1.06	2.22	1.23	1.24	1.26	0.10	15.06	0.14	0.38	0.79	21.72*	3.82	41.93*	1.22
Sweden	2.51*	1.82	1.22	0.66	3.25*	0.10	17.17*	0.46	1.08	0.56	7.05	0.46	49.5*	2.34

Note: *The countries which have higher average than the overall average of the EU countries.

Chapter 2

Spatial Regression Model Specification and Measurement Errors

Anıl Eralp[*,†,§] **and Rukiye Dağalp**[‡,¶]

†*Department of Econometrics, Bolu Abant İzzet Baysal University, Bolu, Turkey*

‡*Department of Statistics, Ankara University, Ankara, Turkey*

§*anil.eralp@ibu.edu.tr*

¶*rdagalp@ankara.edu.tr*

There are various models that allow the spatial autocorrelation between observations to be considered in modeling. However, it is seen that the effects of measurement errors in both the response and explanatory-variables on the unknown parameters of these models have not been adequately investigated. In this study, within the framework of spatial linear mixed model and spatial econometric models (spatial lag model and spatial error model), the effects of measurement errors in the existing literature are discussed theoretically and taken into account that both or only one of the response and explanatory variables are measured with error.

---

*Corresponding author.

1.   Introduction

In many articles in the literature, the covariates with measurement errors have been studied in the context of observed data. However, it is seen that the covariates with measurement error are handled in relatively few studies concerning spatial models (Huque *et al.,* 2016; Suesse, 2018). The rise in the use of spatial data analysis further increases the importance of measurement errors on model specification. So, in this study, various regression specifications using spatial data with the classical measurement error models are theoretically examined. In the literature, it was found that there is no consensus in this context. Therefore, the advantages and disadvantages of various model specification approaches is discussed from different perspectives.

In conventional regression models using cross-sectional datasets, observations for a variable are assumed to be independent of each other. Due to the observations being obtained from points in space or regional locations, spatial dependence is often encountered between observations. Spatial dependence is that observation values in a location tend to be similar to observation values in nearby locations; spatial regression methods are used to explain the cause of this dependence (LeSage, 2008).

The results of statistical analysis are reliable, useful and meaningful, depending on the measurements of the data that is used. Sometimes the substitute is observed instead of the variable that is actually desired to be observed. In other words, the difference between the true value of observation and its observed values gives the measurement error. In such cases, the true variable X said to be error-free cannot be accurately measured, and the substitute variable X^* said to be error-prone is observed with a measurement error U instead of the real variable X. This substitution of X^* for X creates problems in the analysis of the data, usually referred to as measurement error problems. The random variable X^* can be defined by the classical error model $X^* = X + U$ and it is assumed that the measurement error U is a random variable with a distribution of $N(0, \sigma_U^2)$ and unrelated to X (Carroll *et al.,* 2006). On fitting a regression model with error-prone covariates, the effects of measurement error causes inconsistencies in parameter estimation and also its statistical inferences (Stefanski, 1985; Fuller, 1987). Nevertheless, in practice, there is an opinion that measurement errors are small enough to be neglected and can be ignored in the analysis. In contrast, the view that measurement errors can be ignored by studies in this area has become less prevalent in recent years.

In applied studies in many areas, it is inevitable to encounter measurement errors. Therefore, measurement errors are often encountered in spatially defined covariates. Failure to take measurement errors into account in regression specifications reduces the success of naive estimates of regression coefficients. Furthermore, it is suggested that the definition of the classical measurement error cannot be applied in the context of spatial modeling due to the presence of spatial correlation between observations (Huque *et al.*, 2016). In this context, various approaches, such as spatio-temporal model (e.g. Carroll *et al.*, 1997; Xia and Carlin, 1998), spatial linear mixed model (e.g. Li *et al.*, 2009; Huque *et al.*, 2014), semiparametric model (e.g. Huque *et al.*, 2016) and spatial econometrics model with measurement error (e.g. Suesse, 2018), have been proposed as the model specification in the literature. In this context, the effects of measurement errors in the existing literature are theoretically discussed within the framework of spatial linear mixed model and spatial lag model and spatial error model from spatial econometric models.

2. Measurement Errors

For a meaningful statistical inference, good measurement and appropriate statistical analysis are essential. The statistical inference for real-world problems must be based on true and good measurements. Unfortunately, it is not easy to reach the true variable and measurement quality of a variable has many practical consequences and difficulties. Measurement uncertainty of random variables gives rise to a variety of complications and causes too many unexpected problems in the statistical inference which is based on correctly measured observations. Presence of measurement error in random variables has been extensively studied in many areas especially focused on regression analysis. Regression analysis is a statistical methodology adopted to reveal the relationship between the response and explanatory variables which are assumed to be observed without any error. In this regard, the basic assumption is violated in the presence of measurement error, especially in the explanatory variables. In applications, variables are either too expensive, unavailable for measurement or mismeasured. There are several ways of obtaining and defining the variables. Certain variables are not directly measurable, such as intelligence, ability and climatic conditions. Thus, a variable should be identified error-prone variable either unmeasureable, so-called latent variable, or measured with error, so-called error-prone variable which is the sum of the true value and the measurement error. Although some variables, such as nutrient intake, systolic blood pressure, radiation dose and radon gas, are clearly defined, it can be difficult to obtain correct observations. A variable could be conceptually well defined, but it may not be possible to obtain correct observations, e.g. experience measured in number of years. Even if a variable is quite understandable, it can be a qualitative variable by its nature. As an example, intelligence quotient (IQ) scores can be given. In fact, it is not possible to prevent measurement error, especially in economic variables. As sources of measurement errors in economic variables, registration errors and actually unobservable variables of interest can be given (Maddala, 2001). In all these cases, the true observation of a variable cannot be directly measured, rather the variable is observed with some error. The observed variable in place of the true variable is called as measurement error or errors-in-variables or error-prone variable. This substitution of the true variable creates problems in the analysis of data, generally referred to as measurement error problems, and the statistical models used to analyze such data are called *measurement error models* or *errors-in-variables models*. In practice, measurement error problems occur in many areas, such as environmental, agricultural or medical investigations and economic analyses. The problem of measurement errors is one of the most fundamental problems in statistical analysis whose theories depend on error-free observations (Gokmen *et al.*, 2020). The presence of measurement errors causes biased and inconsistent parameter estimates and leads to erroneous conclusions to various degrees in statistical analysis (Carroll *et al.*, 2006).

In traditional regression analysis, the response variable Y is a dependent variable (regressand) whose variation can be explained by an explanatory variable or independent variable X (regressor) which must be observable. Sometimes, the explanatory variable X or the response variable Y cannot be observed since measurement of these variables are difficult because it is either too expensive, impossible owing to unavailability, or prone to mismeasurement. The situation in which the measurement error has the most impact in statistical analysis is when the explanatory variable is observed with error (Carroll *et al.*, 2006; Fuller, 1987). For this reason, the explanatory variable is mostly taken into account as an error-prone variable in regression analysis. The measurement error models can be defined, according to the characteristic of X, as either *functional models* with fixed X or *structural models* with random X (Fuller, 1987). In the literature, structural models are generally considered in measurement error models.

Consider the classical linear regression model with one explanatory variable that is unobservable

$$Y = \alpha + \beta X + \varepsilon \tag{1}$$

and a classical measurement error model

$$X^* = X + U \tag{2}$$

where the true predictor X is measured with error as X^* which is a surrogate for X and U is the measurement error independent of X^* (Carroll *et al.*, 2006; Maddala, 2001). If X^* is assumed to be a surrogate for X, then U and Y are conditionally independent given X, so that the error U and the residual $Y - E(Y|X)$ are uncorrelated and the residual is also uncorrelated with X. The measurement error U could be homoscedastic (constant variance) or heteroscedastic.

Regression analysis is structured on the conditional distribution of the response variable Y given the explanatory variable X. If the conditional distribution of Y given $(X; X^*)$ is the same as the conditional distribution of Y given X, that is $f_{Y|X,X^*} = f_{Y|X}$, the error in X^* as a measurement of X is called nondifferential, and X^* is said to be a *surrogate* for X. If $f_{Y|X,X^*} \neq f_{Y|X}$, the error in X^* is called differential (Carroll *et al.*, 2006; Fuller, 1987). In most of the statistical methods in the literature, the presence of the measurement error is based on the assumption that X^* is taken as a surrogate. In this situation, it is not possible to carry out a direct statistical analysis of the dependence of Y on X if a substitute variable X^* is observed instead of the true variable X.

In this study, the classical measurement error model with known measurement error variance is considered and X^* is taken as a surrogate for X. When fitting the regression model to the observed data, the parameter estimators obtained by ignoring the measurement error are referred to as *naive* estimators which are usually biased and inconsistent estimators of the true parameters in the regression of Y given X. The naive estimators of the regression coefficients of variables measured with error are biased toward zero. This type of bias is called *attenuation* or *attenuation toward the null*, which is easy to understand and well known in the context of simple linear regression. The amount of attenuation is called the *reliability ratio* (Fuller, 1987) and is commonly denoted by λ. The reliability ratio provides an approximate measure of attenuation which is the proportion of the variation of error-free variable in the variation of error-prone variable.

Consider the multivariate normal formulation of the simple linear regression model in Eq. (1):

$$\begin{pmatrix} Y \\ X \end{pmatrix} \sim N \left(\begin{pmatrix} \alpha + \beta_X \mu_X \\ \mu_X \end{pmatrix}, \begin{pmatrix} \beta_X^2 \sigma_X^2 + \sigma_\varepsilon^2 & \beta_X^2 \sigma_X^2 \\ \beta_X^2 \sigma_X^2 & \sigma_X^2 \end{pmatrix} \right)$$

If X^* is observed instead of the true explanatory variable X, the substitute variable X^* is jointly normally distributed with $(Y; X)$, then the multivariate normal model for $(Y; X; X^*)$ is

$$\begin{pmatrix} Y \\ X \\ X^* \end{pmatrix} \sim N \left(\begin{pmatrix} \alpha + \beta_X \mu_X \\ \mu_X \\ \mu_{X^*} \end{pmatrix}, \right.$$

$$\left. \begin{pmatrix} \beta_X^2 \sigma_X^2 + \sigma_\varepsilon^2 & \beta_X^2 \sigma_X^2 & \beta_X \sigma_{XX^*} + \sigma_{\varepsilon X^*} \\ \beta_X^2 \sigma_X^2 & \sigma_X^2 & \sigma_{XX^*} \\ \beta_X \sigma_{XX^*} & \sigma_{XX^*} & \sigma_{X^*}^2 \end{pmatrix} \right)$$

where the variances of X, X^* and ε are σ_X^2, $\sigma_{X^*}^2$ and σ_ε^2, respectively. The covariances are $\sigma_{XX^*} = \mathrm{Cov}(X, X^*)$ and $\sigma_{\varepsilon X^*} = \mathrm{Cov}(\varepsilon, X^*)$. The parameter estimations of regression model on the data (Y, X) and (Y, X^*) are called true and naive estimators, respectively. Consistent estimating equation of intercept α and slope β based on the error-free data, given by the likelihood function, is

$$\sum_{i=1}^{n}(Y_i - \alpha - \beta X_i)\begin{pmatrix}1\\X_i\end{pmatrix} = \begin{pmatrix}0\\0\end{pmatrix} \tag{3}$$

Equation (3) yields the ordinary least squares (LS) estimator for slope, which is given by

$$\hat{\beta}_{\text{True}} = \frac{\sum_{i=1}^{n}(Y_i - \bar{Y})(X_i - \bar{X})}{\sum_{i=1}^{n}(X_i - \bar{X})^2}$$

The ordinary LS estimator for slope on substituting X^* for X in Eq. (3) is given by

$$\hat{\beta}_{\text{Naive}} = \frac{\sum_{i=1}^{n}(Y_i - \bar{Y})(X_i^* - \bar{X}^*)}{\sum_{i=1}^{n}(X_i^* - \bar{X}^*)^2} = \frac{S_{YX^*}}{S_{X^*}^2} = \frac{S_{YX} + S_{YU}}{S_X^2 + 2S_{XU} + S_U^2}$$

where $S_X^2 = S_{XX}$ is the sample variance of $X_1, \ldots, X_n$, $S_U^2 = S_{UU}$ is the sample variance of $U_1, \ldots, U_n$, $S_{X^*}^2 = S_{X^*X^*}$ is the sample variance of $X_1^*, \ldots, X_n^*$. Similarly, S_{YX}, S_{YU} and S_{XU} are the corresponding sample covariances of $(Y_i, X_i), (Y_i, U_i)$ and (X_i, U_i), $i = 1, \ldots, n$. By the law of large numbers, both S_{YU} and S_{XU} converge to zero, $S_X^2 \xrightarrow{\text{P}} \sigma_X^2$ and $S_U^2 \xrightarrow{\text{P}} \sigma_U^2$ as $n \to \infty$. On account of this,

$$\hat{\beta}_{\text{Naive}} = \frac{S_{YX^*}}{S_{X^*}^2} \xrightarrow{\text{P}} \lambda\beta, \quad \text{as n} \to \infty$$

where $\lambda = \sigma_X^2/(\sigma_X^2 + \sigma_U^2)$ is the reliability ratio and $\beta = \sigma_{YX}/\sigma_X^2$ is the true regression slope. When σ_X^2 is positive and finite, the attenuation factor λ is a real number in the range $[0;1]$. It is seen that $\hat{\beta}_{\text{Naive}}$ is biased and attenuated toward zero since the inequality $\beta_{\text{Naive}} < \beta_{\text{True}}$ shows that the naive estimator of slope is always biased toward zero. The extreme cases are

$\lambda = 1 \Leftrightarrow \sigma_U^2 = 0$, no measurement error and

$\lambda = 0 \Leftrightarrow \sigma_U^2 \to \infty$, data are all noise due to measurement error

Otherwise, the naive analysis results in a biased estimate of the true slope (Fuller, 1987).

The effects of measurement error can be examined in three different situations which are only the explanatory variable, or only the response variable or both the explanatory and response variables observed with measurement error.

Case 1: When X^* is observed for true regressor X, according to Eq. (1), the regression model is given as

$$\left.\begin{array}{l} Y = \alpha + \beta X + \varepsilon \\ X^* = X + U_1 \end{array}\right\} \Rightarrow$$

$$Y = \alpha + \beta X + \varepsilon = \alpha + \beta(X^* - U_1) + \varepsilon = \alpha + \beta X^* + \underbrace{\varepsilon - \beta U_1}_{\varepsilon_1}$$

$$= \alpha + \beta X^* + \varepsilon_1$$

where U_1 is the measurement error with $N(0, \sigma_{U_1}^2)$. The covariance of the surrogate and the model error term is

$$\mathrm{Cov}(X^*, \varepsilon_1) = \mathrm{Cov}(X + U_1, \varepsilon - \beta U_1) = -\beta \sigma_{U_1}^2$$

which is a violation of the assumption that the explanatory variable and model error term are uncorrelated.

The attenuation induced by measurement error on the regressor is illustrated in Fig. 1. For this illustration, the data were generated with $\alpha = 12.90, \beta = 0.70$ and a sample size of 100. The random variables are generated from the normal distribution with $(X, U, \varepsilon)^T \sim N\{(0, 0, 0)^T, \mathrm{diag}(1, 1, 0.5)\}$.

Case 2: When Y^* is observed for the true response variable Y, according to Eq. (1), the regression model is given as

$$\left.\begin{array}{l} Y = \alpha + \beta X + \varepsilon \\ Y^* = Y + U_2 \end{array}\right\} \Rightarrow Y^* - U_2 = \alpha + \beta X + \varepsilon$$

$$Y^* = \alpha + \beta X + \underbrace{\varepsilon - \beta U_2}_{\varepsilon_2} = \alpha + \beta X + \varepsilon_2$$

where U_2 is the measurement error with $N(0, \sigma_{U_2}^2)$. The covariance of the explanatory variable and model error term is

$$\mathrm{Cov}(X, \varepsilon_2) = \mathrm{Cov}(X, \varepsilon + U_2) = 0$$

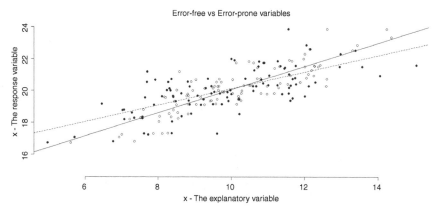

Fig. 1. Illustration of the simple linear regression model with a classical measurement error model. The steeper line and empty circles are the least squares fit and plot of the true (Y, X) data, respectively. The attenuated line and filled circles are the least squares fit and plot of the observed (Y, X^*) data, respectively.

Thus, the assumption pertaining to the explanatory variable and model error term is satisfied. When the response variable is observed with measurement error, the slope estimator of the regression models is not much affected from the measurement error, which is illustrated in Fig. 2.

Case 3: When Y^* and X^* are observed for the true response variable Y and regressor X, according to Eqs. (1) and (2), the regression model is given as

$$\left. \begin{aligned} Y &= \alpha + \beta X + \varepsilon \\ X^* &= X + U_1 \\ Y^* &= Y + U_2 \end{aligned} \right\}$$

$$Y^* - U_2 = \alpha + \beta(X^* - U_1) + \varepsilon$$
$$\Rightarrow Y^* = \alpha + \beta X^* + \underbrace{\varepsilon - \beta U_1 + U_2}_{\varepsilon_2} = \alpha + \beta X^* + \varepsilon_3$$

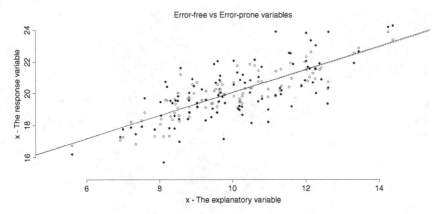

Fig. 2. Illustration of the simple linear regression model with a classical mea-surement error model. The steeper line and empty circles are the least squares fit and plot of the true (Y, X) data, respectively. The attenuated line and filled circles are the least squares fit and plot of the observed (Y^*, X) data, respectively.

where U_1 and U_2 are independent measurement errors with

$$N\{(0, 0)^T, \text{diag}\{\sigma_{U_1}^2, \sigma_{U_2}^2\}\}$$

The covariance of the explanatory variable and the model error term is $\text{Cov}(X^*, \varepsilon_3) = \text{Cov}(X + U_1, \varepsilon - \beta U_1 + U_2) = -\beta \sigma_{U_1}^2$, which is a violation of the assumption that the explanatory vari-able and model error term are uncorrelated which is illustrated in Fig. 3.

As a result, the effect of measurement error on regression analysis is generally associated with attenuation, specifically to the classi-cal error model. In reality, the effect of measurement error causes bias which depends critically on the importance of correct identi-fication of the measurement error. Erroneously specified measure-ment error component of a model can create as great a problem as that caused by ignoring the measurement error in statistical analysis (Buzas *et al.*, 2014). Comparing all three cases, the obvious attenu-ation of the slope in simple linear regression model occurs in Cases 1

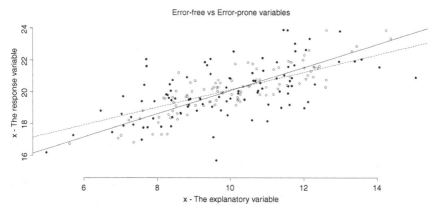

Fig. 3. Illustration of the simple linear regression model with a classical measurement error model. The steeper line and empty circles are the least squares fit and plot of the true (Y, X) data, respectively. The attenuated line and filled circles are the least squares fit and plot of the observed (Y^*, X^*) data, respectively.

Table 1. Slopes and residual variances of the simple regression model relating Y to X^*.

Error Model	Slope	Residual Variance
Differential	$\beta\left(\frac{\sigma_{XX^*}}{\sigma_{X^*}^2}\right) + \left(\frac{\sigma_{\varepsilon X^*}}{\sigma_{X^*}^2}\right)$	$\sigma_\varepsilon^2 + \beta^2 \sigma_X^2 - \frac{(\beta\sigma_{XX^*} + \sigma_{\varepsilon X^*})^2}{\sigma_{X^*}^2}$
Surrogate (Nondifferential)	$\beta\left(\frac{\sigma_{XX^*}}{\sigma_{X^*}^2}\right)$	$\sigma_\varepsilon^2 + \beta^2 \sigma_X^2 (1 - \rho_{XX^*}^2)$
Classical	$\beta\left(\frac{\sigma_{XX^*}}{\sigma_X^2 + \sigma_U^2}\right)$	$\sigma_\varepsilon^2 + \beta^2 \sigma_X^2 \left(\frac{\sigma_U^2}{\sigma_X^2 + \sigma_U^2}\right)$
No measurement error	β	σ_ε^2

Note: X^* is a differential measurement, a surrogate or an unbiased classical-error measurement, and the case of no error.
Source: (Buzas *et al.*, 2004).

and 3, where the explanatory variable is taken as a surrogate (Fig. 3). The effect of measurement error on slope coefficient and variance of errors under different conditions is given in Table 1.

3.   Spatial Regression Models

The classical linear regression model in matrix notation is given as follows:

$$y = X\beta + e$$

where y is an $n \times 1$ dimensional vector containing the observation values of the response variable, X is a non-stochastic $n \times k$ dimensional matrix containing the observation values of the explanatory variable y, β is a vector of coefficients with dimension of $k \times 1$ and e is the $n \times 1$ dimensional vector of random error terms.

In the classical regression model, the mean of the error terms is zero, so $E[e] = 0$.[1] They also have the same and independent distribution (*i.i.d.*), i.e. $E[ee^T] = \sigma^2 I^2$ and there is no correlation between the error terms. In such a generation data process, it is argued that even if the observations consist of "regional" units, the observation units are independent of each other.

Although the assumption of independence between observations is really important for theory and greatly simplifies the classical regression model, this simplification is unlikely to be appropriate due to the possibility of spatial dependence between the error terms in the context of spatial data. If the residuals or explanatory variables are spatially dependent, the results of the predicted model will be biased and inconsistent. Spatial dependence means that the values of nearby locations are influenced and possess similar attributes. In other words, regional observations have generally interrelated and characteristically similar features to each other (Fischer and Wang, 2011; Legendre and Legendre, 1998).

3.1.   *Spatial linear mixed model*

A linear mixed model (LMM) is a linear parametric model that includes the fixed effects parameters associated with the term or categorical explanatory variables and the random effects of one or

---

[1]In $E[e] = 0$,0 is a vector of $n \times 1$ zeros.
[2]I is the $n \times n$ dimensional unit matrix.

more random factors. In this model, fixed effects show the effects of the explanatory variable, while random effects show the effects of clusters and levels. Thus, the random effects can be used directly to model the alterations in the response variable at different levels (West *et al.*, 2007). Therefore, possible clustering in spatial data can be modelled with an LMM.

An LMM is expressed as

$$Y = X\beta + Z\gamma + e$$

where β is the vector of unknown fixed effects parameters and γ is the vector of unknown random effects parameters. e is the error term with a distribution $N(0, \sigma_e^2 I)$. Also, $\gamma \sim N(0, \sigma_\gamma^2 I)$ and $\text{Cov}(\gamma, e) = 0$ are assumed. Under this LMM, the variance of Y is

$$V = Z(\sigma_\gamma^2 I)Z^T + \sigma_e^2 I$$

and the generalized least squares (GLS) estimator is obtained as (Huque *et al.*, 2014; Zhang *et al.*, 2005)

$$\hat{\beta} = (X^T \hat{V}^{-1} X)^{-1} X^T \hat{V}^{-1} Y$$

3.2. *Spatial econometrics models*

Consider the general spatial model as follows:

$$y = \rho W_1 y + X\beta + e$$
$$e = \delta W_2 e + \varepsilon \tag{4}$$

where e and ε are error terms, in which $\varepsilon \sim N(0, \sigma_\varepsilon^2)$, ρ and λ are spatial autocorrelation coefficients and W_1 and W_2 are weight matrices.

Based on the study of Anselin (1988), spatial dependence can be defined in two basic ways with different constraints on Eq. (4). These are spatial lag dependence and spatial residual correlation. The models are based on economic theories acting as a spread effect under the interaction between neighboring locations or points in space. Therefore, spatial lag models explain the spatial correlation (dependency) in the response variable.

The spatial lag model or spatial autoregressive model (SAR) is obtained as an extension of the classical linear regression model with the response variable, weighted with the weight matrix W_1, taken as an explanatory variable. Spatial lag model is provided by the general spatial model with the constrain $\delta = 0$ as follows:

$$y = \rho W y + X\beta + e \qquad (5)$$

where ρ is the coefficient of spatial autocorrelation (a measurement of spread effect), W is the weight matrix and e is the random error term assumed to be $N(0, \sigma_e^2 I)$. The reduced form of the model in Eq. (5) is

$$y = A^{-1}X\beta + A^{-1}e \qquad (6)$$

where $A = I - \rho W$. The maximum likelihood estimation method can be used to estimate the unknown coefficients of the model in Eq. (6) because of the nonlinearity of the model.

Since the conditional distribution of y depends on the distribution of error terms e, the log-likelihood function based on y can be written as follows:

$$\ln(L) = -\frac{n}{2}\ln(2\pi\sigma_e^2) + \ln|A| - \frac{1}{2\sigma_e^2}(Ay - X\beta)^T(Ay - X\beta) \qquad (7)$$

Taking the partial derivatives of the log-likelihood function in Eq. (7) with respect to the unknown parameters vector, the maximum likelihood estimators of the unknown coefficient β vector, variance $(\hat{\sigma}_e^2)$ and variance–covariance matrix $(\hat{\Omega})$ of the error term are obtained as

$$\hat{\beta} = (X^T X)^{-1} X^T A y$$

$$\hat{\sigma}_e^2 = \frac{1}{n} E(e^T e) \qquad (8)$$

$$\hat{\Omega} = \sigma_e^2 (A^T A)^{-1}$$

According to the spatial lag model, the spatial error model (SEM) takes into account the spatial dependence of the error terms. For example, spatial error dependence may arise from spatially correlated unobservable latent variables. Furthermore, it can also emerge due to variables collected for analysis from neighbors that do not fully reflect field boundaries (Fischer and Wang, 2011).

The SEM can be represented as a spatial lag model as in Eq. (4) with the autoregressive process arising from the spatial correlation between the error terms of the classical linear regression model with the constraint $\rho = 0$ on a general spatial model given as

$$y = X\beta + e \qquad (9)$$

$$e = \lambda We + \varepsilon \qquad (10)$$

where e is the random error term of the regression model as in Eq. (9), which is also an autoregressive process as in Eq. (10) under the assumption of being spatially autocorrelated and ε is the random error term assumed to be $N(0, \hat{\sigma}_\varepsilon^2 I)$. The model in Eq. (10) can be written as

$$e = B^{-1}\varepsilon$$

where $B = I - \lambda W$. In this situation, the models in Eqs. (9) and (10) can be revised as

$$y = X\beta + B^{-1}\varepsilon \qquad (11)$$

The log-likelihood function based on y in Eq. (11) can be written as follows:

$$\ln(L) = -\frac{n}{2}\ln(2\pi\sigma_\varepsilon^2) + \ln|B| - \frac{1}{2\sigma_\varepsilon^2}(By - BX\beta)^T(By - BX\beta)$$

$$(12)$$

Taking the partial derivatives of log-likelihood function in Eq. 12) with respect to the unknown parameters vector, the maximum likelihood estimators of the unknown coefficient β vector, variance $(\hat{\sigma}_\varepsilon^2)$ and variance–covariance matrix $(\hat{\Omega})$ of the error term are obtained as (Anselin, 1988; LeSage and Pace, 2007; 2009).

$$\hat{\beta} = (X^T X)^{-1} X^T y$$

$$\hat{\sigma}_e^2 = \frac{1}{n} E(\varepsilon^T \varepsilon)$$

$$\hat{\Omega} = \sigma_\varepsilon^2 (B^T B)^{-1}$$

4.   Measurement Errors in Spatial Regression Models

There are a limited number of studies examining the effects of measurement error on the estimators of unknown parameters in spatial regression models based on spatial data. It is seen that these very few studies focus on the explanatory variable measured with error, generally called the classical measurement error. This may be because, as mentioned before, the coefficients of the regression model have a greater effect in the presence of measurement error in the explanatory variable (see Fig. 2).

As for spatial regression models based on spatial data, (Suesse, 2018) is the only known study that examines the effects of measurement errors on spatial econometric models obtained by adding the effect of spatial dependence based on spatial data to the model as an external information. In this study, the SAR and SEM models take into consideration the presence of measurement error in the explanatory variable.[3]

In this section, the effect on the estimates of the coefficients in the SAR and SEM models from the spatial econometric models in the presence of measurement error in the explanatory variable is investigated (Suesse, 2018).

4.1.   *Measurement errors in the response variable* Y[4]

The simple linear regression model with one explanatory variable can be written as

$$Y_i = \alpha + \beta X_i + e_i \tag{13}$$

where i indicates the spatial location points. Y_i^* denotes unobserved real values of the response variable. Y_i is observed values taken as a

---

[3]In addition, Le Gallo and Fingleton (2012) has examined by considering the error term and can show a spatial structure in examining the effect of measurement error on parameter estimation in the explanatory variable observed as a surrogate variable. At this point, the investigated model can be considered as the SEM model.

[4]Suesse (2018) was referred to in the writing of this section.

substitute of the true response variable and can be given as

$$Y_i^* = Y_i + u_i \tag{14}$$

and considering the measurement error in Eq. (13), the model is

$$Y_i - u_i = \alpha + \beta X_i + e_i$$
$$Y_i = \alpha + \beta X_i + v_i \tag{15}$$

As can be seen in the model in Eq. (15), $v_i = u_i + e_i$ is the measurement error model and a composite form of the error term. In the model in Eq. (15), u and e are the error terms with zero mean and if the explanatory variable X, u and e are accepted to be pairwise uncorrelated to each other, the LS estimator of unknown coefficient vector β is an unbiased estimator given as

$$\hat{\beta} = (X^T X)^{-1} X^T Y$$

for the model in Eq. (13). That is, the measurement error in the response variable Y does not influence the unbiasedness of the LS estimator, but it affects the estimators of variance and standard deviation of the LS estimator $\hat{\beta}$ as seen in[5]

$$\mathrm{Var}(\hat{\beta}) = \frac{\sigma_u^2 + \sigma_\varepsilon^2}{\sum x_i^2} \tag{16}$$

Since variance is a positive quantity, the variance of $\hat{\beta}$ according to Eq. (16) is estimated to be larger than the actual value in the presence of measurement error in Y.

The probability of spatial dependence among spatial data is quite high. Depending on the source of the spatial dependence, various spatial econometric models can be developed by adding this information to the model. First of all, the effect of measurement error in the response variable will be discussed for the SEM model, created for the case of spatial dependence in the error term, and then for the SAR model.

---

[5]Calculated by the method of deviations from the means.

Assume that Y is observed as the true variable Y^* with a measurement error as in Case 2 in spatial error (Eq. (10)) with $u \sim N(0, \sigma u^2)$ can be written as

$$y = X\beta + v \qquad (17)$$

where the error term is $v = A^{-1}\varepsilon + u$ with $A = I - \rho W$, ρ is a spatial autocorrelation parameter and measures spread effect and W is the weight matrix. The random error terms ε and u are independently normally distributed having zero means and variances σ_ε^2 and σ_u^2, respectively. Therefore, the random error term v has a normal distribution with zero mean and variance $\sigma_v^2 = \sigma_u^2 + \sigma_\varepsilon (A^T A)^{-1}$.

The log-likelihood function for the model in Eq. (17) can be written as

$$\ln(L) = -\frac{n}{2} \ln(2\pi) - \frac{n}{2} \ln \sigma_u^2 + \frac{1}{2} \ln(M) - \frac{1}{2\sigma_u^2} r^T M r \qquad (18)$$

where

$$M = \left(I + \frac{\sigma_\varepsilon^2}{\sigma_u^2} (A^T A) \right)^{-1}$$

$$r = y - X\beta$$

The estimator of the coefficient of the SEM model under measurement error is obtained as

$$\hat{\beta} = (X^T M X)^{-1} X^T M y$$

which is obviously different from the parameter estimator in Eq. (8) and the variance estimator of measurement error is

$$\hat{\sigma}_u^2 = \frac{1}{n} r^T M r$$

In this case, the variance–covariance matrix is

$$\hat{\Omega} = \hat{\sigma}_u^2 M^{-1}$$

Thus, the standard errors of the estimators of the model coefficients will vary depending on the variance of the measurement error (Table 1). In the spatially lag model in Eq. (5), as in the SEM model with measurement errors, the true response variable Y is assumed to be observed as Y^* with a measurement error, as in Eq. (15), and the spatial lag model with $u \sim N(0, \sigma_u^2)$ is given as

$$y = \rho W y + X \beta + v \qquad (19)$$

where the random error term is $v = A^{-1}\varepsilon + u$ with $A = I - \rho W$. As can be seen, the error terms are the same in both the SEM and SAR models with the measurement error. ρ is a spatial autocorrelation parameter and measures spread effect, and W is the weight matrix. The random error terms ε and u are independently normally distributed with zero means and variances σ_ε^2 and σ_u^2, respectively. Thus, the random error term v has a normal distribution with zero mean and variance $\sigma_v^2 = \sigma_u^2 + \sigma_\varepsilon^2 (A^T A)^{-1}$.

The log-likelihood function for the model in Eq. (19) can be written as

$$\ln(L) = -\frac{n}{2}\ln(2\pi) - \frac{n}{2}\ln\sigma_u^2 + \frac{1}{2}\ln(M) - \frac{1}{2\sigma_u^2}r_A^T M r_A$$

where

$$M = \left(I + \frac{\sigma_\varepsilon^2}{\sigma_u^2}(A^T A)\right)^{-1}$$

$$r_A = Ay - X\beta$$

The estimator of the coefficient in the SAR model with measurement error is as follows:

$$\hat{\beta} = (X^T X)^{-1} X^T A y$$

which is the same as the estimator of the coefficient in the model given in Eq. (8) with measurement error. The variance of the measurement error is determined as

$$\hat{\sigma}_u^2 = \frac{1}{n}r_A^T M r_A$$

and the variance–covariance matrix is obtained as

$$\hat{\Omega} = \hat{\sigma}_u^2 M^{-1}$$

In this case, the standard errors of the estimator of the model parameters vary due to the variance of the measurement error (Suesse, 2018).

4.2.   *Measurement errors in the explanatory variable X*[6]

The linear regression model with a single explanatory variable can be written as follows:

$$Y_i = \alpha + \beta X_i + e_i \tag{20}$$

where X_i is the true explanatory variable, i's are spatial locations and $e_i = [e_i, \ldots, e_n]^T$ is the random error variables with $e_i \sim N(0, \sigma_U^2)$. The multivariate normal distribution for the response variable Y and the surrogate variable X^* due to the classical measurement error model in Eq. (2) can be written as follows:

$$\begin{pmatrix} Y \\ X^* \end{pmatrix} \sim \text{MVN} \left(\begin{pmatrix} \alpha + \beta \mu_X \\ \mu_X \end{pmatrix}, \begin{pmatrix} \sigma_e^2 + \beta^2 \sigma_X^2 & \beta \sigma_X^2 \\ \beta \sigma_X^2 & \sigma_X^2 + \sigma_U^2 \end{pmatrix} \right)$$

and the conditional expected value and variance of Y given X^* are

$$E(Y|X^*) = \alpha + \beta(I - \Lambda)\mu_X + \beta\Lambda X^*$$
$$\text{Var}(Y|X^*) = \sigma_e + \beta^2(I - \Lambda)\sigma_X$$

where $\Lambda = \sigma_X^2(\sigma_X^2 + \sigma_U^2)^{-1}$. Obviously, there is a bias due to the variance caused by the error-prone covariate (X^*). If the observed explanatory variable X^* is used in the model in Eq. (20) instead of the true explanatory variable X, the LS estimator becomes

$$\hat{\beta} = (X^{*T}X^*)^{-1}X^{*T}Y$$

For simplicity, when the mean of the variable X is assumed to be zero, the LS estimator converges in probability to

$$\hat{\beta} \xrightarrow{P} \begin{pmatrix} 1 & 0 \\ 0 & (\sigma_X^2 + \sigma_U^2) \end{pmatrix}^{-1} \begin{pmatrix} 1 & 0 \\ 0 & \sigma_X^2 \end{pmatrix} \beta = \begin{pmatrix} 1 & 0 \\ 0 & \lambda \end{pmatrix} \beta$$

where $\lambda = \frac{\sigma_X^2}{\sigma_X^2 + \sigma_U^2}$ is the reliability ratio (attenuation factor). Moreover, the attenuation factor can be written if the diagonal of the σ_X

---

[6]Huque *et al.* (2014) was referred to in the writing of this section.

matrix consists of constant σ_X^2 and the diagonal of the σ_U matrix consists of fixed σ_U^2. Therefore, the GLS estimator is

$$
\hat{\beta}^{gls} = \begin{pmatrix} n & 0 \\ 0 & (\sigma_a^2)^{-1}(\sigma_X^2 + \sigma_U^2) \end{pmatrix}^{-1} \begin{pmatrix} n & 0 \\ 0 & (\sigma_a^2)^{-1} + \sigma_X^2 \end{pmatrix} \beta
$$

$$
= \begin{pmatrix} 1 & 0 \\ 0 & \lambda^{gls} \end{pmatrix} \beta
$$

where

$$
\lambda^{gls} = \frac{1 + (\sigma_a^2)^{-1}\sigma_U^2}{(\sigma_a^2)^{-1}\sigma_X^2}
$$

is the reliability ratio (Huque *et al.*, 2014).

5. Conclusion and Discussion

In spatial regression models based on spatial data, the measurement error has not been sufficiently investigated in terms of both response and explanatory variables. In the literature, there has not been any research on the presence of measurement error in terms of spatial econometric models, except for the measurement error in the response variable. Therefore, the measurement error in the explanatory variable for spatial econometric models emerges as a subject to be examined.

There are a limited number of studies examining the effects on the parameter estimations in the presence of measurement error in spatial regression models based on spatial data. These studies seem to focus on the measurement error in the explanatory variable, so-called the classical measurement error. On account of this, the effect of the measurement error in the explanatory variable on the parameter estimation is greater than that of the measurement error in the response variable, as mentioned earlier (see Fig. 2).

The effects of measurement errors on unknown parameter estimates are not affected by spatial effects. In terms of non-spatial cross-sectional data, measurement error in the explanatory variable causes a bias toward zero and inconsistent parameter estimates. On the other hand, the measurement error in the response variable has relatively less effect on parameter estimations. It is seen that the effects of measurement error in the explanatory variable in spatial lag and spatial error models have not been investigated yet. Suesse (2018) reported that the effect of measurement error on the parameter estimate can be minimized by estimating the parameter via maximum likelihood method for large data under the assumption that the effects of measurement error in the response variable are quite small. However, the study indicates that regarding the effects of the magnitude of the measurement error on the SEM model, the estimates of the parameters are biased toward zero under the constant spatial dependency effect. As a result, the effects of measurement error on model prediction apart from spatial data should be investigated in case of the presence of measurement error in the variables. In addition, it is necessary to investigate the effects of measurement errors in explanatory variables in spatial econometric models.

References

Anselin, L. (1988). *Spatial Econometrics: Methods and Models*. Springer.

Buzas, J.S., Stefanski, L.A., and Tosteson, T.D. (2014). *Measurement Error*. In: Ahrens, W. and Pigeot, I. (eds.), Handbook of Epidemiology. Springer, New York, NY. https://doi.org/10.1007/978-0-387-09834-0_19.

Carroll, R.J., Ruppert, D., Stefanski, L.A., and Crainiceanu, C.M. (2006). *Measurement Error in Nonlinear Models: A Modern Perspective*. CRC Press.

Carroll, R.J., Chen, R., George, E., Li, T., Newton, H., Schmiediche, H., and Wang, N. (1997). Ozone exposure and population density in harris county, Texas. *Journal of the American Statistical Association*, **92**(438): 392–404.

Fischer, M.M. and Wang, J. (2011). Spatial data analysis: models, methods and techniques. Springer Science & Business Media.

Fuller, W.A. (1987). *Measurement Error Models*. John Wiley & Sons.

Gokmen, S., Dagalp, R., and Kilickaplan, S. (2020). Multicollinearity in measurement error models. *Communications in Statistics-Theory and Methods*, 1–12.

Huque, M.H., Bondell, H.D., and Ryan, L. (2014). On the impact of covariate measurement error on spatial regression modelling. *Environmetrics*, **25**(8): 560–570.

Huque, M.H., Bondell, H.D., Carroll, R.J., and Ryan, L.M. (2016). Spatial regression with covariate measurement error: A semiparametric approach. *Biometrics*, **72**(3): 678–686.

Le Gallo, J. and Fingleton, B. (2012). Measurement errors in a spatial context. *Regional Science and Urban Economics*, **42**(1–2): 114–125.

Legendre, P. and Legendre, L. (1998). *Numerical Ecology*. 2nd Edn. Elsevier, Amsterdam.

LeSage, J.P. (2008). An introduction to spatial econometrics. *Revue d'économie industrielle*, **123**(3e): 19–44.

LeSage, J.P. and Pace, R.K. (2007). A matrix exponential spatial specification. *Journal of Econometrics*, **140**(1): 190–214.

LeSage, J.P. and Pace, R.K. (2009). *Introduction to Spatial Econometrics*. CRC Press, Boca Raton.

Li, Y., Tang, H., and Lin, X. (2009). Spatial linear mixed models with covariate measurement errors. *Statistica Sinica*, **19**(3): 1077–1093.

Maddala, G.S. (2001). *Introduction to Econometrics*. Wiley.

Stefanski, L.A. (1985). The effects of measurement error on parameter estimation. *Biometrika*, **72**(3): 583–592.

Suesse, T. (2018). Estimation of spatial autoregressive models with measurement error for large data sets. *Computational Statistics*, **33**(4): 1627–1648.

West, B.T., Welch, K.B., and Galecki, A.T. (2007). *Linear Mixed Models: A Practical Guide using Statistical Software*. CRC Press, Boca Raton.

Xia, H. and Carlin, B.P. (1998). Spatio-temporal models with errors in covariates: Mapping Ohio lung cancer mortality. *Statistics in Medicine*, **17**(18): 2025–2043.

Zhang, L., Gove, J.H., and Heath, L.S. (2005). Spatial residual analysis of six modeling techniques. *Ecological Modelling*, **186**(2): 154–177.

Chapter 3

Determining Harmonic Fluctuations in Food Inflation

Yılmaz Akdi[†,‖], Kamil Demirberk Ünlü[*,‡,],**
Cem Baş[§,††], and Yunus Emre Karamanoğlu[¶,‡‡]

[†]*Department of Statistics, Ankara University,*
Ankara, Turkey

[‡]*Department of Mathematics, Atılım University,*
Ankara, Turkey

[§]*Price Statistics Directorate, Turkish Statistical Institute,*
Ankara, Turkey

[¶]*Gendarmerie and Coast Guard Academy,*
Ankara, Turkey

[‖]*akdi@ankara.edu.tr*
[**]*demirberk.unlu@atilim.edu.tr*
[††]*cembas@tuik.gov.tr*
[‡‡]*eyunus@bilkent.edu.tr*

In this study, we start with a brief expression of consumer price index of Turkey. In the next step, we give the theoretical essentials of periodogram-based unit root and harmonic regression model. Periodogram-based unit root test is used to identify both the stationarity of data and periodicities. Periodicity is beyond seasonality; it is the hidden cycles in the data. Thus, it is harder to detect them compared to seasonal cycles. Harmonic-regression-type trigonometric regression models are useful in modeling data which have hidden periodicity. Afterward,

---

[*]Corresponding author.

the stationarity properties of monthly inflation and monthly food inflation of Turkey for the period between 2004 and 2020 are investigated.
Standard augmented Dickey–Fuller unit root test shows that both series
are integrated of order one. However, the periodogram-based unit root
test shows that monthly inflation has unit root but monthly food inflation does not. After examining the unit root, the hidden cycles in the
food inflation are revealed. The cycles in food inflation are important
because they may trigger a headline inflation. The main contribution of
this study is the identification of the hidden cycles in food inflation. It
has cycles of approximately two, four, six and eight years. These cycles,
in short, correspond to cycles of two years of consecutive periods.

1.   Introduction and Preliminaries

Inflation refers to continuous and general increase which is observed
in prices of goods and services in the economy. Inflation is a corner-
stone not only for individuals and firms but also for the whole economy. Inflation deeply affects decision-making phase of consumers and
investors. Continuous and volatile rise experienced in prices leads to
the comparison of different goods and services by the consumers and
makes selection of products difficult. This causes consumers to limit
their consumption and firms to delay their investments because of
not being able to predict future costs and profits. Interest rates on
which households and firms invest their savings, credit interests used
to purchase goods and services, salaries and wages, social security
premiums and retirement salaries are directly affected by inflation
rate increase. In this scope, consumer price index (CPI) is an important indicator of inflation.

The CPI measures the weighted average price change in goods
which are consumed by households. Calculation of CPI consists of
three steps which are as follows: (i) formation of the consumption
basket; (ii) weighting goods in the basket; (iii) computing the basket's
cost.

The main data source for the determination of goods and services
to be included in the CPI consumption basket and the creation of
weights is the household budget survey (HBS), where household consumption expenditures are determined. According to the European
Statistics Office (Eurostat), goods and services weighing more than
1/1000th of household consumption expenditures are included in the

basket. Instead of HBS, individual consumption expenditures of the institutional population, expenditures made by foreign visitors and administrative records are also used in order to create the consumption basket and weighted goods.

The CPI baskets and weights are updated at the end of each year and chained with the Laspeyres formulation. The index is calculated by dividing current prices to the prices of previous December's price, which is "new price reference period (P_0)," and then chained by multiplying it with the chained index numbers of December's price. The CPI examines prices of goods reflected to the consumer, including transportation and trade margin, and all taxes and product prices are calculated by geometric average.

The main expenditure group with the highest weight in the CPI consumption basket is "Food and soft drinks," in line with the "hierarchy of needs" concept Maslow defined in the 1950s. The weight of this main spending group, which is essential for meeting the physiological needs of each individual regardless of the socio-economic situation, decreased slightly over time due to changes in household consumption patterns and increase of weight in other consumption groups; this weight is 22.77% for 2020. Food and soft drinks comprise 134(32.1%) of 418 products in the CPI basket. Prices of vegetables and fruits and other food in the CPI consumption basket are collected to reflect the price fluctuations in the market every week, and it is important that the definitions of items are the same in order to follow prices properly. The weights of the main expenditure groups in the CPI in the last five years are given in Table 1. The impact of food groups on inflation in terms of both level and volatility is explained by periodic movements in food prices and higher weight of food prices in the consumer basket of Turkey.

The two most important reasons why inflation has recently remained above the target and highly volatile are the reaction of prices to exchange rate movements and increases in food prices. It is difficult to explain the increase in food prices only through climate conditions or agricultural policy. A wide range of negative factors, from market failures and irrigation infrastructure to limited producer's organizations lead to volatile food prices. The fact that food consumptions are in the first place among the mandatory consumption needs is the reason they have the highest weight among

Table 1.   Weights of the groups in CPI.

Groups	Year				
	2016	2017	2018	2019	2020
1. Food and soft drinks	23.68	21.77	23.03	23.29	22.77
2. Alcoholic drinks and tobacco	4.98	5.87	5.14	4.23	6.06
3. Clothing and shoes	7.43	7.33	7.21	7.24	6.96
4. Housing	15.93	14.85	14.85	15.16	14.34
5. Furniture	8.02	7.72	7.66	8.33	7.77
6. Health	2.66	2.63	2.64	2.58	2.80
7. Transport	14.31	16.31	17.47	16.78	15.62
8. Communication	4.42	4.12	3.91	3.69	3.80
9. Entertainment and culture	3.81	3.62	3.39	3.29	3.26
10. Education	2.56	2.69	2.67	2.40	2.58
11. Restaurant and hotels	7.47	8.05	7.27	7.86	8.67
12. Other goods and services	4.73	5.04	4.76	5.15	5.37

the main expenditure groups in the CPI. Therefore, correct analysis of the food sector and especially the food prices are very critical for the fight against inflation (TSB, 2018).

The main aim of this study is to investigate the monthly inflation and monthly food inflation by using periodogram-based time series analysis. Recently, Akdi *et al.* (2020a, 2020b, 2020c) and Okkaoğlu *et al.* (2020) have shown the power and usefulness of periodogram-based methodology to identify the hidden cycles in energy, climate and macroeconomic data. We start our analysis by investigating the stationarity of the time series, which is followed by identification of the hidden cycles in the data, and finally, we use this information to predict the monthly inflation rates. Our results show that the monthly food inflation for Turkey has cycles of two years which confirms the work of Furceri *et al.* (2016). Moreover, we have showed that although the monthly inflation rates are not stationary, the monthly food inflation rate is stationary. The remainder of this chapter is organized as follows. Literature review is given in Section 2. In Section 3, we introduce the methodology. Section 4 is devoted to the presentation of empirical evidence. Finally, in Section 5, we conclude the study.

2. Literature Review

We have divided the literature into three parts. The first part is devoted to studies on food inflation. The second part is related to recent inflation forecasting literature. The last part is entirely related to the inflation and food inflation research on Turkey.

Tule *et al.* (2019) works on the predictability of agricultural commodity prices in the inflation of Nigeria. They use 12 major agricultural commodities to forecast both food inflation and headline inflation by using various methodologies. Their results show that the agricultural commodities predict both food and headline inflation better than random walk model. The random walk model is the benchmark for this study. Bhattacharya and Sen Gupta (2018a) studies the food inflation of India, and their results reveal that food inflation influences both non-food inflation and headline inflation. Bhattacharya and Sen Gupta (2018b) analyse and determine the food inflation of India for the period between 2006 and 2013. Their results express a pass-through effect from food inflation to non-food inflation and headline inflation. Iddrisu and Alagidede (2020) investigate the relation between monetary policies and food inflation by using quantile regression model. They show that restrictive monetary policy destabilizes food prices in South Africa. Bhattacharya and Jain (2020) use panel vector autoregression to investigate association between food inflation and monetary policy of advanced and emerging economies. They show that unexpected monetary tightening in emerging and advanced economies has a positive effect on food inflation. To the best of our knowledge, the only study that investigates the periodicities of commodity food price is by Rezitis and Sassi (2013). They use structural time series analysis for the monthly dataset spanning the period 1992–2012. Their results reveal a two-year periodicity in the investigated dataset.

Ülke *et al.* (2018) compare the time series analysis and machine learning methodology in forecasting inflation. They use the inflation of the United States (US), and their results indicate that the investigated machine learning algorithms have higher accuracy. Gil-Alana and Mudida (2017) deal with the inflation of Kenya and show evidence of a structural break in the investigated dataset. Sari *et al.* (2016) use backpropagation neural network to forecast interest rate and CPI of Indonesia. They compare the proposed model with Sugeno fuzzy inference system. Their results show that the proposed model performs better than the benchmark. Moshiri and Cameron (2000) compare the performance of backpropagation neural network and econometric models to forecast inflation rates. Their results indicate that in some cases, the proposed model is better than the classical model and in some other cases, they are the same. Baciu (2015) uses stochastic models to forecast inflation rate of Romania. They compare autoregressive, moving average process and composite models. They conclude that each model has its own advantages. Tang and Zhou (2015) compare the optimization techniques for support vector machine (SVM) to predict the inflation rate of China. They conclude that optimization of SVM by particle swarm optimization leads to the best performance. Aparicio and Bertolotto (2020) use online price index to forecast inflation rate of Australia, Canada, France, Germany, Greece, Ireland, Italy, the Netherlands, the United Kingdom and the US by using autoregression, Philips curve and random walk. Their results show that the models used outperform the Survey of Professional Forecasts. Furceri *et al.* (2016) investigate the effect of food inflation on domestic inflation. Their results show that a 10% increase in food inflation causes a 0.5% increase in the domestic inflation.

Önder (2004) evaluates the performance of Philips curve in forecasting inflation of Turkey. The performance of the Philips curve is compared with autoregressive integrated moving average (ARIMA), vector autoregression, vector error correction and a naive no-change model. They show that Philips curve has the best performance. Kara and Orak (2008) introduce the main line with the inflation-targeting regime and based on this framework after the 2001 crisis, Turkey's aims to shed light on the inflation-targeting experience. The study first takes the basic elements that form the framework of inflation targeting, and second, it summarizes the problems specific to developing countries. Finally, implicit inflation-targeting in Turkey is assessed for the period 2002–2005 and is considered under the spotlight of formal inflation-targeting period after 2006, and the final part of all these assessments are carried out within the framework of general conclusions. In addition to a theoretical analysis supported by simulation-based insights, Mahir (2010) reviewed inflation measurement techniques to achieve more accurate indexing for price changes.

Erdem (2017) investigates the causal relationship between food inflation and inflation uncertainty for the Turkish economy. The data
used in the study are monthly and cover the period of 2005–2017
for Turkey's economy. In this study, Kalman filter technique, among
various algorithmic approaches, was used to get inflation uncertainty.
According to the results, there is a one-way causality from food inflation to inflation uncertainty for the Turkish economy. However, there
is no causality from inflation uncertainty to food inflation. Ersin
(2017) aims to determine food accessibility situation of Turkey and
regional differences in terms of food accessibility. The study also aims
to develop a model to identify relations between food accessibility
and food price index. Developing food accessibility index, data from
2006 to 2015 are used for food accessibility indicators of each region.
"Food consumption as a share of household expenditure," "proportion of poverty" and "gross domestic product per-capita" are chosen
as food accessibility indicators, and data for each indicator are normalized to calculate the index. It is revealed that regions getting high
scores of the index are all industrialized regions with higher urban
population.

3. Methodology

Periodograms are often used to detect hidden periodicities in the
stationary time series. Therefore, the time series should be stationary. In the literature, there are many unit root tests, but augmented
Dickey–Fuller (ADF) (Dickey and Fuller, 1979) stands out from the
rest. ADF was developed based on the asymptotic distribution of the
least squares estimators of the parameters. Also, Akdi and Dickey
(1998) suggest a unit root test which stands on periodograms. Periodograms derived by transformation of trigonometric functions are
the most commonly known periodic functions. Let e_t be an independent and identically distributed error term with zero mean and σ^2
variance, then consider the following model:

$$Y_t = \mu + R\cos(wt + \phi) + e_t, \quad t = 1, 2, 3, \ldots, n \qquad (1)$$

Here, μ, R, ϕ and w represent expected value, amplitude, phase and
frequency, respectively. These parameters should be estimated. If w_k
is chosen as $2\pi k/n$, then w_k becomes a Fourier frequency. If we let

$a = R\cos(\phi)$ and $b = R\sin(\phi)$ and use the properties of trigonometric functions, the model in Eq. (1) can be written as

$$Y_t = \mu + a\cos(w_k t) + b\sin(w_k t) + e_t, \quad t = 1, 2, 3, \ldots, n \quad (2)$$

The least squares estimators of the model in Eq. (2) are calculated as $\hat{\mu} = \bar{Y}_n$, $a_k = \frac{2}{n}\sum_{t=1}^{n}(Y_t - \bar{Y}_n)\cos(w_k t)$ and $b_k = \frac{2}{n}\sum_{t=1}^{n}(Y_t - \bar{Y}_n)\sin(w_k t)$.

The calculated a_k and b_k are known as Fourier coefficients. By using the following property of trigonometric functions, we can say that the Fourier frequencies are invariant according to mean:

$$\sum_{t=1}^{n}\cos(w_k t) = \sum_{t=1}^{n}\sin(w_k t) = 0 \quad (3)$$

By using these Fourier coefficients, the periodogram ordinate at w_k frequency of the time series is calculated as

$$I_n(w_k) = \frac{n}{2}(a_k^2 + b_k^2) \quad (4)$$

On the other hand, if $f(w_k)$ represents the spectral density function of the stationary time series $\{Y_1, Y_2, Y_3, \ldots, Y_n\}$, then the normalized periodogram ordinate is asymptoticly distributed as chi-square with two degrees of freedom (Fuller, 1996; Wei, 2006), that is

$$\frac{I_n(w_k)}{f(w_k)} \xrightarrow{D} \chi_1^2 \quad \text{as } n \to \infty$$

Although periodograms are usually used to identify the hidden periodicities of the time series, they are also used to test for unit root. Therefore, Akdi and Dickey (1998) define the following statistics:

$$T_n(w_k) = \frac{2(1 - \cos(w_k))}{\sigma_n^2} I_n(w_k) \quad (5)$$

Under the assumption that the series has a unit root, the distribution of the test statistics can be defined as

$$T_n(w_k) \xrightarrow{D} Z_1^2 + 3Z_2^2 \quad \text{as } n \to \infty \quad (6)$$

Z_1 and Z_2 have standard normal distribution. The distribution of the test statistics can be re-written as

$$T_n(w_k) \xrightarrow{D} \chi_1^2 + 3\chi_2^2 \quad \text{as } n \to \infty \tag{7}$$

An asymptotic distribution is valid for all k values, but it is mostly used for $k = 1$. If the value of the test statistics, let's call it $t_n(w_k)$, is less than a critical value, then the null hypothesis that the series has unit root is rejected. The critical values can be found in Akdi and Dickey (1998). Here, we will list some of the critical values:

$$P(T \leq 0.034818) = 0.01$$
$$P(T \leq 0.178496) = 0.05$$
$$P(T \leq 0.3690089) = 0.10$$

According to the above critical values, if $t_n(w_k) < c_{0.01}$, then the time series is stationary under 1% significance level.

Periodogram-based unit root test offers many advantages. Some of them can be listed as follows: (i) Periodograms are calculated based on the transformation of trigonometric functions. They do not need any model, so they are invariant with respect to model. Also, periodograms are invariant with respect to mean. (ii) Critical values do not depend on the volume of the dataset. They can be used for a small sample size. (iii) Periodogram-based unit root test only needs the parameter estimation of white noises' variance. (iv) Under the assumption of stationarity, the asymptotic distribution of periodograms is chi-square with two degrees of freedom, thus they have analytic power function for the test statistics. (v) If the data have a periodic component, then the unit root test becomes more robust. For this reason, we also utilize the periodogram-based unit root test.

As we mentioned before, periodograms are used to investigate hidden periodicities of the time series. Consider Eq. (2); if the null hypothesis H_0: $a = b = 0$ is rejected, then we can say that the time series has a periodic component. To test this hypothesis, standard F-test is not applicable because frequencies w_k are not known (Wei, 2006). For this hypothesis, the following test statistic is defined (Wei, 2006):

$$V = I_n(w_{(1)}) \left[\sum_{k=1}^{m} I_n(w_{(k)}) \right]^{-1} \tag{8}$$

In the above statistics, $I_n(w_{(1)})$ represents the highest periodogram value, while m is equal to the integer part of $n/2$. Under the null hypothesis of

$$H_0: a = b = 0$$

$$P(V > c_\alpha) = \alpha \cong m(1 - c_\alpha)^{m-1} \tag{9}$$

where c_α represents the critical value with α significance level. By using Eq. (8), the critical values can be found as

$$c_\alpha = 1 - (\alpha/m)^{1/(m-1)} \tag{10}$$

If the value of V is greater than the critical value, that is $V > c_\alpha$, we say that the null hypothesis $H_0: a = b = 0$ is rejected. Hence, the time series has a periodic component. If we modify Eq. (7), we can use the resulting statistics to test other periodicities in the series. To achieve this, we define

$$V_i = I_n(w_{(i)}) \left[\sum_{k=1}^{m} I_n(w_{(k)}) - \sum_{k=1}^{i-1} I_n(w_{(k)}) \right]^{-1} \tag{11}$$

Again, if the values of the above statistics are greater than the critical values, then we say that the corresponding periodicities are statistically significant. In Eq. (11), $I_n(w_{(i)})$ represents the ith highest periodogram value.

Now, assume that some periodic components are detected, namely, p_1 and p_2, then consider the following harmonic regression model:

$$Y_t = \mu_1 + A_1 \cos\left(\frac{2\pi t}{p_1}\right) + B_1 \sin\left(\frac{2\pi t}{p_1}\right)$$

$$+ A_2 \cos\left(\frac{2\pi t}{p_1}\right) + B_2 \sin\left(\frac{2\pi t}{p_1}\right) + e_t, \quad t = 1, 2, 3, \dots, n \tag{12}$$

This model considers the periodic components of the time series, and for this reason, if the time series contains periodicity, it is more appropriate to use this model for forecasting. On the other hand, since the time series is stationary, student-t test can be used to estimate the parameters.

4. Experimental Setup and Results

In this study, monthly inflation rates and monthly food inflation rates between January 2004 and July 2020 are used. The main aim of this section is to detect the periodicity of the investigated time series; afterward, this information is used to utilize harmonic regression model to forecast the future values.

In order to investigate the periodicity of the time series, at first stationarity should be checked. Time series graphs of the two series with their autocorrelation functions (ACF) and partial autocorrelation functions (PACF) are given in Fig. 1.

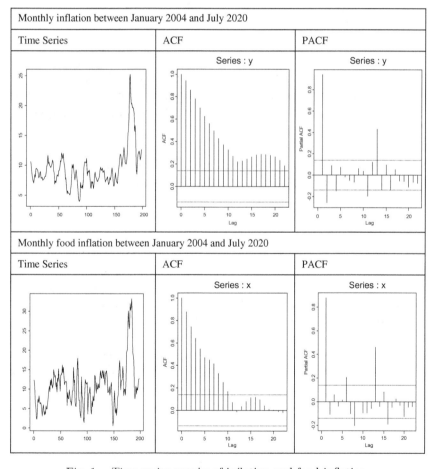

Fig. 1. Time series graphs of inflation and food inflation.

Table 2.   Monthly average of inflation and food inflation rates.

Month	Average Inflation	Average Food Inflation
January	9.589	10.956
February	9.512	10.923
March	9.425	10.728
April	9.419	10.862
May	9.438	10.778
June	9.500	10.741
July	9.419	10.660
August	9.466	10.700
September	9.475	10.629
October	9.555	10.674
November	9.469	10.446
December	9.480	10.571

According to Fig. 1, it can be seen that the monthly food inflation is greater than the monthly inflation between January 2004 and July 2020. The monthly averages of these two inflation rates are given in Table 2.

From Table 2, we see that the monthly food inflation is approximately 10% to 15% higher than the monthly inflation rate. Also, it is known that the percentage of the food inflation in the monthly basket inflation rate is high.

When the ACF of the inflation rate is investigated from Fig. 1, it is seen that the declaration rate is slow. The stationary of the monthly inflation rate is shady. Still the decline in the ACF of food inflation is relatively faster. Videlicet, the monthly food inflation could be stationary. Augmented Dickey–Fuller test results for the monthly inflation and monthly food inflation rates are given in Table 3.

According to the results of ADF test, both the series are stationary at the first difference, so we conclude that they are integrated of order one, I(1). Also, we used periodogram-based unit root test since there can be periodicity in the time series. In order to calculate the test statistics, first variances of the white noise of the time series should be estimated. Any consistent estimator is enough to estimate the variance.

In order to determine a suitable model for time series data, different ARIMA models are considered as candidates. The model with

Table 3. Unit root test results.

Test Critical Values	t-Statistics	Probability
Panel A. Monthly Inflation Rate (Level)		
Augmented Dickey–Fuller test statistics	**−1.027**	0.7434
1% level	−3.466	
5% level	−2.877	
10% level	−2.575	
Panel B. Monthly Inflation Rate (First Difference)		
Augmented Dickey–Fuller test statistics	**−8.188**	0.0000
1% level	−3.466	
5% level	−2.877	
10% level	−2.575	
Panel C. Monthly Food Inflation Rate (Level)		
Augmented Dickey–Fuller test statistics	**−2.029**	0.2744
1% level	−3.466	
5% level	−2.877	
10% level	−2.575	
Panel D. Monthly Food Inflation Rate (First Difference)		
Augmented Dickey–Fuller test statistics	**−6.957**	0.0000
1% level	−3.466	
5% level	−2.877	
10% level	−2.575	

the smallest Akaike information criterion (AIC) has been chosen for both monthly inflation and monthly food inflation. The estimation of the white noises is made according to these models. If we denote the monthly inflation by Y_t and let $e_{1,t} \sim WN(0, \sigma^2)$, then the ARIMA model is as follows:

$$Y_t = \alpha_0 + \alpha_1 Y_{t-1} + \alpha_1 Y_{t-1} + \alpha_{11} Y_{t-11} + \alpha_{12} Y_{t-12}$$
$$+ \alpha_{13} Y_{t-13} + e_{1,t}, \quad t = 1, 2, 3, \ldots, n$$

which has an AIC of 536.6463, and in the same manner, if the monthly food inflation is denoted by X_t and $e_{2,t} \sim WN(0, \sigma^2)$, the

Table 4. Periodogram based unit root test results.

	Monthly Inflation	Monthly Food Inflation
$I_n(w_1)$	614.0250	415.1790
$\hat{\sigma}_n^2$	0.854382	5.319356
$t_n(w_1)$	0.723650	0.078591

Table 5. Periods of monthly food inflation.

i	$I_n(w_{(1)})$	V_i	Period	5% Critical Value	Test Result
1	1789.841	0.279	66 months	0.074	Significant
2	1026.003	0.222	49.5 months	0.074	Significant
3	510.759	0.142	99 months	0.074	Significant
4	415.179	0.135	198 months	0.074	Significant
5	333.520	0.125	24.75 months	0.074	Significant

model for this case is as follows:

$$X_t = \beta_0 + \beta_1 X_{t-1} + \beta_6 Y_{t-6} + \beta_8 Y_{t-8} + \beta_{12} Y_{t-12}$$
$$+ \beta_{13} Y_{t-13} + e_{2,t}, \qquad t = 1, 2, 3, \ldots, n$$

The model of the monthly food inflation has an AIC of 898.7364. According to these models and Eq. (5), the variance of the white noise parameter estimation are given in Table 4.

According to the results in Table 4, the value of test statistics for the monthly inflation data is calculated as $t_n(w_1) = 0.72365$. This statistic is greater than the critical value of 10% significance level, so the monthly inflation data have unit root. In the case of monthly food inflation, the calculated test statistics is $t_n(w_1) = 0.078591$ which is less than the critical value of 1% significance level. Monthly food inflation is stationary under 1% significance level.

According to these results, periodic components of the monthly food inflation can be investigated. The highest five periodogram values and the corresponding test statistics with their periods are given in Table 5. The critical values which calculated by Eq. (9) are given in Table 6.

Tables 5 and 6 indicate that the monthly food inflation has periods of 24.75 months (approximately two years), 198 months (the whole

Table 6.   Critical values of V_i.

α	0.01	0.02	0.03	0.04	0.05
c_α	0.089	0.083	0.079	0.077	0.074
α	0.06	0.07	0.08	0.09	0.10
c_α	0.073	0.071	0.070	0.069	0.068

period of the dataset), 99 months (half of the period of the dataset), 49.5 months (approximately four years) and 66 months (5.5 years). It does not make sense to consider the 198-month period which includes the whole period of the dataset. In short, we can say that there are two-year cycles in the data.

Considering the identified periodicities of the data, we estimate the harmonic regression model given in Eq. (12) as

$$X_t = \mu + A_1 \cos\left(\frac{2\pi t}{66}\right) + B_1 \sin\left(\frac{2\pi t}{66}\right) + A_2 \cos\left(\frac{2\pi t}{49.5}\right)$$

$$+ B_2 \sin\left(\frac{2\pi t}{49.5}\right) + A_3 \cos\left(\frac{2\pi t}{99}\right) + B_3 \sin\left(\frac{2\pi t}{99}\right)$$

$$+ A_4 \cos\left(\frac{2\pi t}{198}\right) + B_4 \sin\left(\frac{2\pi t}{198}\right) + A_5 \cos\left(\frac{2\pi t}{2475}\right)$$

$$+ B_5 \sin\left(\frac{2\pi t}{24.75}\right) + e_t, \quad t = 1, 2, \ldots, 198 \tag{13}$$

where $e_t \sim WN(0, \sigma^2)$. The parameter estimations with their significance levels are given in Table 7.

The investigated time series data are stationary, so to test the significance of the parameters, t-test is used. For the harmonic regression given in Eq. (13) H_0: $A_1 = 0$ and H_0: $A_3 = 0$, null hypothesis cannot be rejected. In light of this information, we re-estimate the harmonic regression model for the monthly food inflation as

$$X_t = \mu + B_1 \sin\left(\frac{2\pi t}{66}\right) + A_2 \cos\left(\frac{2\pi t}{49.5}\right) + B_2 \sin\left(\frac{2\pi t}{49.5}\right)$$

$$+ B_3 \sin\left(\frac{2\pi t}{99}\right) + A_4 \cos\left(\frac{2\pi t}{198}\right) + B_4 \sin\left(\frac{2\pi t}{198}\right)$$

$$+ A_5 \cos\left(\frac{2\pi t}{24.75}\right) + B_5 \sin\left(\frac{2\pi t}{24.75}\right) + e_t,$$

$$t = 1, 2, \ldots, 198 \tag{14}$$

The parameter estimations with their significance levels are given in Table 8.

Also, we have forecasted the future values of monthly food inflation from July 2020 to April 2024 by using the model in Eq. (14). Forecasted values are given in Table 9.

Table 7. Parameter estimation of the harmonic regression model (Eq. (13)).

Variable	DF	Parameter Estimate	Standard Error	t-Value	$P > |t|$
μ	1	10.72551	0.25107	42.72	<0.0001
A_1	1	0.27679	0.35506	0.78	0.4366
B_1	1	−4.24295	0.35506	−11.95	<0.0001
A_2	1	−1.58403	0.35506	−4.46	<0.0001
B_2	1	−2.80259	0.35506	−7.89	<0.0001
A_3	1	−0.47051	0.35506	−1.33	0.1867
B_3	1	−2.22212	0.35506	−6.26	<0.0001
A_4	1	1.59353	0.35506	4.49	<0.0001
B_4	1	−1.28622	0.35506	−3.62	0.004
A_5	1	−0.70288	0.35506	−1.98	0.0492
B_5	1	1.69554	0.35506	4.78	<0.0001

Table 8. Parameter estimation of the harmonic regression model (Eq. (14)).

Variable	DF	Parameter Estimate	Standard Error	t-Value	$P > |t|$
μ	1	10.72551	0.25131	42.68	<0.0001
B_1	1	−4.24295	0.35540	−11.94	<0.0001
A_2	1	−1.58403	0.35540	−4.46	<0.0001
B_2	1	−2.80259	0.35540	−7.89	<0.0001
B_3	1	−2.22212	0.35540	−6.25	<0.0001
A_4	1	1.59353	0.35540	4.48	<0.0001
B_4	1	−1.28622	0.35540	−3.62	0.0004
A_5	1	−0.70288	0.35540	−1.98	0.0494
B_5	1	1.69554	0.35540	4.77	<0.0001

Table 9. Forecasted values by using model in Eq. (14).

Year	Month	Forecasted Value	Year	Month	Forecasted Value	Year	Month	Forecasted Value
2020	July	9.5526	2022	January	2.6584	2023	July	13.7318
2020	August	9.1230	2022	February	2.9954	2023	August	13.4771
2020	September	8.7228	2022	March	3.4989	2023	September	13.1495
2020	October	8.3317	2022	April	4.1582	2023	October	12.7792
2020	November	7.9313	2022	May	4.9552	2023	November	12.3969
2020	December	7.5068	2022	June	5.8644	2023	December	12.0313
2021	January	7.0482	2022	July	6.8551	2024	January	11.7077
2021	February	6.5513	2022	August	7.8922	2024	February	11.4462
2021	March	6.0185	2022	September	8.9384	2024	March	11.2610
2021	April	5.4583	2022	October	9.9562	2024	April	11.1590
2021	May	4.8855	2022	November	10.9096	2024	May	11.1401
2021	June	4.3200	2022	December	11.7665	2024	June	11.1974
2021	July	3.7856	2023	January	12.4998	2024	July	11.3174
2021	August	3.3089	2023	February	13.0894	2024	August	11.4816
2021	September	2.9167	2023	March	13.5229	2024	September	11.6678
2021	October	2.6351	2023	April	13.7960	2024	October	11.8513
2021	November	2.4869	2023	May	13.9128	2024	November	12.0077
2021	December	2.4906	2023	June	13.8851	2024	December	12.1137

5.   Conclusion

In this study we aimed to identify the hidden periodic structure of the investigated time series monthly inflation rate and monthly food inflation rate. Periodicity differs from seasonality. It is beyond seasonality; it is the hidden cycles found in the time series. For this reason, classical time series methods, such as ARIMA, cannot capture this component of the data. Thus, we use the model given in Eq. (14) to capture the periodicity. One of the findings of our results is that, with accordance to ADF test and periodogram-based unit root test, the monthly inflation rate has unit root but the monthly food inflation rate does not. Thus, we planned to investigate the reason behind the unit root of monthly inflation in depth. We can identify which component of the monthly inflation given in Table 1 causes this nonstationary behaviour and the reasons for this behaviour. In order to investigate the periodicity, the dataset should be stationary. Since the food inflation data are stationary, we worked on this dataset. Our results show that in general, food inflation has cycles of two years.

In this study, we found that food inflation has cycles of approximately two, four, six and eight years. It consists of consecutive two-year cycles. Some reasons behind this two-year cycle can be the following. The planting and production of food in Turkey does not follow any plan and schedule. Farmers have the opportunity to decide to produce based on prices or not. Also, agriculture in Turkey became externally dependent on products, such as diesel, electricity, fertilizers, oilseeds and corn. Imports were expensive as a result of the increase in exchange rates, while the costs to the farmers increased. Fluctuation in prices continues without resolving structural problems, such as irrigation infrastructure and generalization of cooperatives. Moreover, if certain products are scarce in a certain year, the government may export these goods. Thus, in the following year, farmers may not plant these exported goods, which causes an increase in the prices. Lack of a general and comprehensive production plan causes these two-year cycles. A two-year cycle implies that the price of goods in the inflation basket has hidden price changes that happens in two-year periods. Further, we identified the dynamics of the food inflation by decomposing it into groups.

References

Akdi, Y. and Dickey, D.A. (1998). Periodograms of unit root time series: Distributions and tests. *Communications in Statistics—Theory and Methods*, **27**(1): 69–87.

Akdi, Y., Gölveren, E. and Okkaoğlu, Y. (2020a). Daily electrical energy consumption: Periodicity, harmonic regression method and forecasting. *Energy*, **191**: 116524.

Akdi, Y., Varlik, S. and Berument, M.H. (2020b). Duration of global financial cycles. *Physica A: Statistical Mechanics and its Applications*, 124331.

Akdi, Y., Okkaoğlu, Y., Gölveren, E. and Yücel, M.E. (2020c). Estimation and forecasting of PM 10 air pollution in Ankara via time series and harmonic regressions. *International Journal of Environmental Science and Technology*, **17**(8): 3677–3690.

Aparicio, D. and Bertolotto, M.I. (2020). Forecasting inflation with online prices. *International Journal of Forecasting*, **36**(2): 232–247.

Baciu, I.C. (2015). Stochastic models for forecasting inflation rate. Empirical evidence from Romania. *Procedia Economics and Finance*, **20**: 44–52.

Bhattacharya, R. and Jain, R. (2020). Can monetary policy stabilise food inflation? Evidence from advanced and emerging economies. *Economic Modelling*, **89**: 122–141.

Bhattacharya, R. and Sen Gupta, A. (2018b). Drivers and impact of food inflation in India. *Macroeconomics and Finance in Emerging Market Economies*, **11**(2): 146–168.

Dağdur, E. (2017). *Gıda Fiyatları Endeksinin Gıdanın Erişilebilirliğine Etkisi: Türkiye Örneği*, Master Thesis (Ankara University).

Dickey, D.A. and Fuller, W.A. (1979). Distribution of the estimators for autoregressive time series with a unit root. *Journal of the American Statistical Association*, **74**: 427–431.

Erdem, H.F. (2017). Gıda Enflasyonunun Enflasyon Belirsizliği Üzerine Etkisi, Karadeniz Teknik Üniversitesi Sosyal Bilimler Enstitüsü Sosyal Bilimler Dergisi, **7**(14): 425–436.

Erkan, M. (2010). *Inflation Measurement Methods, Criteria and Consequences of Incorrect Indices — Turkey Case*. Master Thesis (Marmara University).

Fuller, W.A. (1996). *Introduction to Statistical Time Series* (New York: Wiley).

Furceri, D., Loungani, P., Simon, J. and Wachter, S.M. (2016). Global food prices and domestic inflation: Some cross-country evidence. *Oxford Economic Papers*, **68**(3): 665–687.

Gil-Alana, L.A. and Mudida, R. (2017). CPI and inflation in Kenya. Structural breaks, non-linearities and dependence. *International Economics*, **150**: 72–79.

Iddrisu, A.A. and Alagidede, I.P. (2020). Monetary policy and food inflation in South Africa: A quantile regression analysis. *Food Policy*, **91**: 101816.

Kara, A.H. and Orak, M. (2008). Gıda Fiyatları Endeksinin Gıdanın Erişilebilirliğine Etkisi: Türkiye Örneği, İstanbul Ekonomik Tartışmalar Konferansı.

Moshiri, S. and Cameron, N. (2000). Neural network versus econometric models in forecasting inflation. *Journal of Forecasting*, **19**(3): 201–217.

Okkaoğlu, Y., Akdi, Y. and Ünlü, K.D. (2020). Daily PM10, periodicity and harmonic regression model: The case of London. *Atmospheric Environment*, 117755.

Önder, A.Ö. (2004). Forecasting inflation in emerging markets by using the Phillips curve and alternative time series models. *Emerging Markets Finance and Trade*, **40**(2): 71–82.

Rezitis, A.N. and Sassi, M. (2013). Commodity food prices: Review and empirics. *Economics Research International*.

Sari, N.R., Mahmudy, W.F. and Wibawa, A. (2016). Backpropagation on neural network method for inflation rate forecasting in Indonesia. *International Journal of Soft Computing and its Applications*.

Tang, Y. and Zhou, J. (2015). The performance of PSO-SVM in inflation forecasting. In *2015 12th International Conference on Service Systems and Service Management (ICSSSM)*, (IEEE).

TSB. (2018). Turkish Presidency of Strategy and Budget website. http://www.sbb.gov.tr/wp-content/uploads/2020/04/EnflasyonlaMucadele OzelIhtisasKomisyonuRaporu.pdf. Accessed 19 June 2020.

Tule, M.K., Salisu, A.A. and Chiemeke, C.C. (2019). Can agricultural commodity prices predict Nigeria's inflation? *Journal of Commodity Markets*, **16**: 100087.

Ülke, V., Sahin, A. and Subasi, A. (2018). A comparison of time series and machine learning models for inflation forecasting: Empirical evidence from the USA. *Neural Computing and Applications*, **30**(5): 1519–1527.

Wei, W.W. (2006). *Time Series Analysis — Univariate and Multivariate Method*. (Boston: Pearson Addison Wesley).

Chapter 4

Nonlinear and Chaotic Time Series Analysis

Baki Ünal[*,†,§] **and Çağdaş Hakan Aladağ**[‡,¶]

†*Department of Industrial Engineering,*
İskenderun Technical University, Hatay, Turkey
‡*Department of Statistics, Hacettepe University, Ankara,*
Turkey

§*baki.unal@iste.edu.tr*
¶*aladag@hacettepe.edu.tr*

This study introduces the techniques used in the analysis of nonlinear and deterministic time series. Since chaotic time series are nonlinear and deterministic, the techniques introduced here can be successfully applied to the analysis of chaotic time series. Chaotic time series differ from stochastic time series. Therefore, the analysis of both types of time series shows differences. In this work, for analyzing nonlinear deterministic time series, we present how the phase space is reconstructed, how chaos is detected, how noise is cleaned from the data and how the deterministic time series are estimated.

1. Introduction

Chaos is a new paradigm in science. This paradigm has led to new perspectives and understandings in many scientific fields. Chaotic

---

[*]Corresponding author.

systems can be expressed precisely with differential equations or difference equations, but it is not possible to predict such systems in the long run due to sensitivity to initial conditions. In chaotic systems, a small difference in initial conditions can lead to large differences in final outcomes. This phenomenon is called the butterfly effect. Although chaotic systems cannot be predicted in the long run, they can be predicted in the short run. A method used for short-term prediction of chaotic systems is presented in this chapter. Since chaotic systems are nonlinear systems, these systems can be analyzed with nonlinear time series analysis (TSA) methods. The foundations of these methods were developed in the 1980s by Packard *et al.* (1980) and Takens (1981). During these years, deterministic chaos became a popular research area and researchers began to investigate chaos in systems in the laboratory and in nature. These methods involve reconstructing the phase space to obtain delayed vectors using a one-dimensional time series data, as described later. By these methods, quantities such as Lyapunov exponents and fractal dimensions can be calculated, near future state of the time series can be predicted and in some cases, even equations describing dynamics can be obtained from the time series. However, nonlinear TSA methods also have some limitations. In this context, the biggest limitation is the noise contained in the observed time series. The negative effects of noise in the time series are presented. Many methods have been proposed in the literature for noise cleaning and noise evaluation. Some of these methods are mentioned here.

The purpose of TSA is to make inferences about the dynamics behind the observed data sequentially over time. The first approaches put forward in this context include linear stochastic models, namely autoregressive (AR) and moving average (MA) models. These stationary Gaussian stochastic processes are characterized by the autocorrelation function.

Like other statistical analyses, TSA includes validation of appropriate data models with certain hypotheses. TSA is essentially a data compression process. In this context, a small number of characteristic numbers is calculated from a big data sample.

The methods known as nonlinear TSA can be highly effective if the data model has deterministic dynamics in a certain phase space. In this analysis framework, information about the hidden unchanging properties of the underlying dynamic system can be provided. In some cases, even equations describing the motion of the system can be obtained. If the underlying system is low-dimensional and deterministic, this analysis framework can reveal the relationships between geometry (fractal dimension), instability (Lyapunov exponents) and unpredictability (entropy).

The most well-known chaotic model in the literature is the chaotic model introduced by Lorenz and defined by the following differential equations:

$$\frac{dx}{dt} = \sigma(y - x) \tag{1}$$

$$\frac{dy}{dt} = x(\rho - z) - y \tag{2}$$

$$\frac{dz}{dt} = xy - \beta z \tag{3}$$

In the equations above, usually the parameters are selected as $\sigma = 10$, $\rho = 28$, $\beta = 8/3$. When this system of equations is solved numerically, the time series and strange attractor in Fig. 1 are obtained.

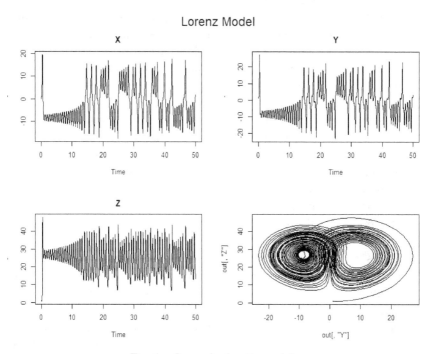

Fig. 1. Lorenz's chaotic model.

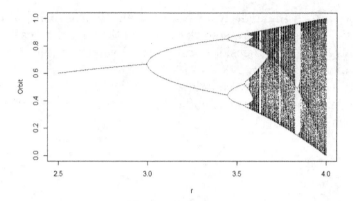

Fig. 2.   Bifurcation diagram of logistic map.

Another well-known discrete chaotic model in the literature is logistic map. Logistic map is a difference equation and expressed as $x_{n+1} = rx_n(1 - x_n)$. Bifurcation diagram of a logistic map is shown in Fig. 2. As seen in this graph, the system changes between periodic and chaotic regimes as the parameter r changes.

There is no consensus on the definition of chaos in the literature, but the most accepted definition of chaos belongs to Devaney (1989, p. 49). Devaney defines chaos with three features.

Suppose X is a metric space. If a continuous function f from X to itself has the following three properties, then f is called chaotic:

1. f is topologically transitive. For all non-empty open sets $U, V \subset X$, there exists a natural number k such that $f^k(U) \cap V$ is non-empty. This condition can be called irreducibility or indecomposability and is a topological condition. This condition implies that a chaotic system cannot be decomposed into two subsystems.
2. Periodic points are dense in space. This condition can be called element of regularity and is a topological condition. This condition means there is order in chaos, such as periodic points.
3. f has sensitive dependence on initial conditions. So, there is a positive constant δ such that for every point x of X and every neighborhood N of x, there exists a point y in N and a nonnegative integer n such that the nth iterates $f^n(x)$ and $f^n(y)$ are more than distance δ apart. This condition implies unpredictability and is not a topological condition but depends on a metric.

This sensitivity is measured by Lyapunov exponents which reveal the average rate of divergence between two adjacent trajectories.

Chaotic systems are characterized with strange attractors. There are two strangeness in a strange attractor. One of them is sensitivity to initial conditions, which is reflected by Lyapunov exponents, and the other one is the attractor's geometry which is reflected by fractal dimensions.

The organization of our work is as follows. In Section 2, we discuss phase space reconstruction method. Phase space reconstruction methods are the initial step in nonlinear and chaotic TSA. In Section 3, we review methods for detecting chaos. These methods are divided into two classes as metric tools and topological tools. The metric tools supply statistics or numeric metrics for judging whether the time series is chaotic or not. These consist of BDS test, correlation dimension, Lyapunov exponent and surrogate data testing. The topological tools are recurrence plot (RP) and recurrence quantification analysis (RQA). RP is a visual tool which must be interpreted by an observer and involves subjectivity. In RQA, the subjectivity of the observer can be suppressed by applying simple pattern recognition algorithms to RP. In Section 4, we discuss some methods to deal with noise in the data. Noise in the data causes adverse effects during analysis. We present several methods to clean noise from the data. In Section 5, we discuss forecasting in nonlinear and chaotic time series. It is impossible to predict a chaotic system in the long run. However, we provide a method for forecasting chaotic systems in the short run. In Section 6, we conclude.

2. Phase Space Reconstruction Methods

Every dynamical system is defined with its phase space. Evolution of a dynamical system can be described in phase space in a geometric form. Every point in the phase space of a dynamical system corresponds to instantaneous state of the dynamical system and represents the entire system. Motion of a dynamical system is defined by trajectories in phase space. These trajectories can be closed curves but cannot have intersections. If there are intersections, there would be many successive points, so evolution of the system cannot be defined uniquely.

When we don't have the equations defining the evolution of a dynamical system, we apply a procedure called embedding to define the evolution of a dynamical system. By utilizing embedding procedure, we can construct multi-dimensional vector series from one-dimensional scalar time series by using fixed time delays.

Suppose we have a one-dimensional scalar time series as follows:

$$\mathbf{x} = (x_1, x_2, x_3, \ldots, x_n) \tag{4}$$

To reconstruct $\mathbf{x}$ by an embedding procedure, two parameters must be determined. These parameters are embedding dimension D and time delay τ. The reconstructed multi-dimensional time series can be demonstrated in a matrix form as follows:

$$\mathbf{V} = \begin{pmatrix} \mathbf{V}_1 \\ \mathbf{V}_2 \\ \vdots \\ \mathbf{V}_{n-(D-1)\tau} \end{pmatrix} = \begin{pmatrix} x_1 & x_{1+\tau} & \cdots & x_{1+(D-1)\tau} \\ x_2 & x_{2+\tau} & \cdots & x_{2+(D-1)\tau} \\ \vdots & \vdots & \ddots & \vdots \\ x_{n+(D-1)\tau} & x_{n+(D-2)\tau} & \cdots & x_n \end{pmatrix} \tag{5}$$

Embedding theorems (Takens, 1981; Mane, 1981) guarantee that geometric object defined by reconstructed multi-dimensional time series is equivalent to the trajectory of original multi-dimensional time series. By this equivalence, features of the original time series, which are called dynamical invariants, are preserved in the reconstructed time series. The important point here is that the reconstructed multi-dimensional time series is obtained from a single measured variable of the dynamical system.

Whitney (1936) proved that it is possible to embed every D-dimensional smooth manifold in a $2D + 1$-dimensional real space. However, Whitney (1936) proved this theorem only for integer dimension values and did not state how likely a given map really forms an embedding. Sauer *et al.* (1991) extended Whitney's theorem to fractal dimensions.

Reconstruction of an attractor of a dynamical system is important. An attractor is defined as a subset of state space of the dynamical system. A time delay embedding is a time-independent map from an attractor to m-dimensional real space ($\mathbb{R}^m$). Takens (1981) proved that a delay map with $m = 2D + 1$ dimension is an embedding of a compact manifold with D dimensions with a smooth measurement function. Sauer *et al.* (1991) generalized the Takens theorem to fractal dimensions and named it "Fractal Delay Embedding Prevalence Theorem." The most important consequence of embedding theorems is that fractal dimension is much more important than the underlying dimension of the true state space.

To determine the correct values of the embedding dimension and time delay, no certain rules are asserted in the literature, but some prescriptions have been suggested. Specifications of embedding parameters depends on the problem in hand. Embedding procedure is generally performed after the transient period has been discarded.

As demonstrated by Takens (1981), to prevent self-intersections of the reconstructed trajectory, the embedding dimension must be greater than twice the attractor dimension. However, this is a sufficient condition but not a necessary condition. Therefore, in practice, the embedding dimension can be chosen lower values.

To determine minimal sufficient embedding dimension D, Kennel *et al.* (1992) suggested a method called false nearest neighbor (FNN). In this method, we start with a small embedding dimension (such as one) and we count the number of neighbors of a point. If we increase embedding dimension by one, some of the previous neighbors will not be neighbors. These are false neighbors. Then, when almost all the neighbors have become real, we can determine the minimum embedding dimension.

In the embedding process, time delay, which is the time between successive elements in the reconstructed vectors, must be specified. Embedding theorems do not say anything about how to determine time delays. However, in practical applications, the time delay must be determined carefully.

If the time delay is chosen too small, there will be no difference between the elements of reconstructed vectors. This phenomenon is named as redundancy by Casdagli *et al.* (1991) and Gibson *et al.* (1992). Furthermore, if the data contain significant noise, it is hard to acquire information from the data. If the time delay is chosen very large, the elements of reconstructed vectors will be independent and uncorrelated. In this situation, even if the real attractor is simple, the reconstructed attractor can be complicated.

There is no theoretical guidance for choosing optimal value of time delay. However, in the literature, several different methods are proposed for choosing time delay. Autocorrelation function of the time series is proposed to determine the time delay. In this sense, the time delay can be chosen, where the autocorrelation function decays to $1/e$. However, this method is grounded on linear correlations and omits nonlinear correlations. To overcome this problem, time-delayed mutual information statistic is proposed to determine the time delay (Fraser and Swinney, 1986). In this context, the first minimum of the time-delayed mutual information is selected as the time delay.

In the literature, there are many proposals to choose the optimal time lag. Liebert and Schuster (1989) proposed "wavering product," Buzug and Pfister (1992) proposed "fill factor" and Rosenstein *et al.* (1994) proposed "displacement from diagonal" statistics to determine the time lag. However, the time lag must be optimized according to a particular application.

However, keep in mind that the usage of FNN and AMI to determine embedding parameters are just prescriptions. In practice, investigations with various embedding parameters can be useful.

If points in the reconstructed phase space are close to each other because of being close in time, and not because of the geometry of the trajectory, then the obtained estimates, which are based on comparison of reconstructed vectors, can be biased if the vectors are not statistically independent. To overcome the effects of this temporal nearness, a condition which imposes that the distance between pairs of points should be greater than the Theiler window is adopted (Theiler, 1986). To estimate the Theiler window, Provenzale *et al.* (1992) suggested to use space–time separation plot. The Theiler window must be set a value after which regular oscillations begin to show up in space–time separation plot.

Embedding procedure can be performed by using principal component analysis (PCA). By utilizing PCA, from a set of variables, the most important variables can be selected. In PCA, the original set is transformed into a new set of uncorrelated variables called principal components. In PCA, the importance of the principal components is measured by standard deviation of each component and proportion of the variance in the data explained by each component.

The embedding dimension can be reduced by using the most important principal components. PCA can also filter noise.

3. Chaos Detection

In practice, we are interested with low-dimensional (low-complexity) chaos. If the chaotic dynamics are high dimensional, it is practically impossible to detect using finite amount of data. In this case, chaotic dynamics and random dynamics are indistinguishable. If the chaotic dynamic is low dimensional, short-term predictability is possible. However, in this case, nonlinear forecasting methods must be utilized instead of linear methods.

3.1. *Metric tools*

3.1.1. *BDS test*

Brock *et al.* (1996) developed a test for independence, which can be implemented to residuals of any time series model. Their test can be seen as a nonlinear version of the Box–Pierce Q statistic utilized in diagnostic checking of ARIMA methods.

The aim of BDS test is to test whether a time series is generated by a chaotic process. This test is widely used in nonlinear dynamics and chaos theory. In addition to the detection of deterministic chaos, the BDS test can also be used to determine goodness of fit of a statistical model.

Brock *et al.* (1996) developed a nonparametric test for nonlinear structure and serial dependence which can be applied to time series. This test is applicable to forecast errors of estimated statistical models. Null hypothesis of BDS test asserts that the time series is generated by an IID process. However, the alternative hypothesis of BDS test is not specified.

To conduct a BDS test, first the time series is fitted to a model given as follows:

$$y_t = f(x_t, b, u_t) \quad \text{or} \quad u_t = g(y_t, x_t, b) \tag{6}$$

Afterward, it is investigated whether the estimated errors are IID. If the formulation of the fitted model is correct, the estimated errors (residuals) will be IID and pass the test. If residuals could not pass the IID test, then the statistical model is not correctly specified.

BDS can be applied to statistical models specified in the following form:

$$y_t = \mu(x_t, a) + \sigma(x_t, b)u_t \tag{7}$$

In the statement above, a and b are parameter vectors. This statement covers GARCH models presented by Engle (1982) and Bollerslev (1986). The BDS test can be used as a model selection and diagnostic tool for the model types as above.

Grassberger and Procaccia (1983) present the concept of correlation integral to measure fractal dimension of a deterministic time series. Correlation integral measures recurrence frequency of temporal patterns in the data. Correlation integral with embedding dimension m is defined as

$$C_{m,n}(\epsilon) = \frac{1}{\binom{n}{2}} \sum_{1 \le s < t \le n} \sum \chi_\epsilon(\|u_s^m - u_t^m\|) \tag{8}$$

$$C_m(\epsilon) = \lim_{n \to \infty} C_{m,n}(\epsilon) \tag{9}$$

If the stochastic process is absolutely regular, the above limit exists. If the process is independent, the following equation holds:

$$C_m(\epsilon) = C_1(\epsilon)^m \tag{10}$$

Theorem. Let $\{u_t\}$ be IID. If $K(\epsilon) > C(\epsilon)^2$,

$$W_{n,T}(\epsilon) = \sqrt{n}\frac{C_{m,n}(\epsilon) - C_1(\epsilon)^m}{\sigma_m(\epsilon)} \tag{11}$$

converges in distribution to $N(0,1)$, where $\sigma_m(\epsilon)$ is the standard sample deviation of $C_{m,n}(\epsilon) - C_1(\epsilon)^m$. $W_{n,T}(\epsilon)$ is called BDS statistic.

The statistic above is distribution free, so no distributional assumption is needed.

BDS statistic is studied with Monte Carlo simulations by Hsieh and LeBaron (1988). They investigated approximation of asymptotic distribution to finite sample distribution. They found that, when ϵ is taken between one-half to two standard deviations of the data, asymptotic distribution is a good approximation for IID data.

Brock *et al.* (1991) demonstrated that BDS test is valid if the following conditions are met:

- The number of data is greater than 500.
- ϵ lies between 0.5σ and 2σ, where σ is the standard deviation of the series.
- The embedding dimension is lower than $n/200$.

However, this BDS test method has some limitations. First, the application of this method needs quite a high number of data. Second, this method does not have a statistical theory for hypothesis testing.

3.1.2. *Correlation dimension*

The dimension of a system defines the minimum number of variables which is required to replicate the system. A fractal is a geometric object which has the same repeating structure in all scales. For a system to be chaotic, its fractal dimension must be fractional and not an integer. In the literature, to calculate fractal dimension, many methods are proposed. Examples of these are the information dimension, Hausdorff dimension, the box-counting dimension and the correlation dimension. These fractal dimension calculation methods do not always yield equivalent measures. The method proposed by Grassberger and Procaccia (1983) is called correlation dimension and has some advantages such as straightforwardness and quick implementation.

Correlation function is defined as follows and reflects the ratio of points separated by a distance smaller than ε:

$$C(N, m, \varepsilon) = \frac{1}{N(N-1)} \sum_{m \leq t \neq s \leq N} \Theta(\varepsilon - \|X_t - X_s\|) \quad \varepsilon > 0 \quad (12)$$

In the statement above, Θ denotes Heaviside function and $\|\cdot\|$ denotes norm operator. The correlation function determines the likelihood that a pair of randomly picked points are separated by a distance less than ε.

Correlation dimension is obtained from correlation sum. To obtain the correlation dimension, it must be determined how $C(N, m, \varepsilon)$

changes as ε changes. If ε is increased, $C(N, m, \varepsilon)$ increases since the number of close points increase. Grassberger and Procaccia (1983) showed that if ε is small, then $C(N, m, \varepsilon)$ increases at rate D_c. As a result, the following approximation is possible:

$$C(N, m, \varepsilon) \approx \varepsilon^{D_c} \qquad (13)$$

Therefore, the correlation dimension can be obtained from the limit

$$D_c = \lim_{\varepsilon \to 0} \frac{\log C(N, m, \varepsilon)}{\log \varepsilon} \qquad (14)$$

The expression above implies that the correlation dimension depends on selected embedding dimension. According to the increase in D_c as the embedding dimension m grows, two different conclusions can be made. Increasing D_c indefinitely as m increases is a sign of a stochastic system. As m increases, if D_c reaches a finite limit, this indicates a chaotic system. The smallest integer greater than the correlation dimension defines the minimum number of variables needed to represent a chaotic attractor.

Since the correlation dimension depends on an embedding process, the choice of embedding parameters is crucial. Unfortunate choices of embedding parameters can cause misleading results. In addition to embedding parameters, the correlation dimension also depends on a sufficiently small number ε and the norm operator. If the data are limited, it is always possible to choose a small ε such that no two points are close to each other.

Brock (1986) showed the cases when correlation function become independent of the norm choice. However, Kugiumtzis (1997) demonstrated that Brock's (1986) results are not valid in the case of short noisy data

Reliability issues also arise from short data. Enormous amount of data is required for reliable models in the case of high-dimensional chaos. Since the correlation dimension does not maintain time ordering of the data, it does not supply information about the process dynamics (Gilmore 1993a, 1993b).

3.1.3. *Lyapunov exponent*

Despite the fact that chaotic systems have predictable temporal evolution, it is difficult to anticipate their long-term behavior. Small changes in the starting conditions of chaotic systems are magnified exponentially. Lyapunov exponent reflects this magnification. If the maximum Lyapunov exponent of a system is positive, this indicates a sensitivity to initial conditions and divergence at exponential rates. Therefore, Lyapunov exponent can be used to detect chaos. The sign of the Lyapunov exponent determines the convergence or divergence of the nearby trajectories. If the system is chaotic, divergence will not continue infinitely but the trajectories stay in a bounded set.

There are two types of methods for calculating the Lyapunov exponent. These are called direct methods and Jacobian methods. The embedding process is required in both methods. The concept behind these approaches is the tracking of two nearby points and computing the rate of separation of trajectories.

Wolf *et al.* (1985) and Rosenestein *et al.* (1994) provide direct techniques that require calculating the divergence rate of initially nearby trajectories.

The Jacobian methods involve the estimation of Jacobians from data. To estimate the Jacobians and Lyapunov exponent, McCaffrey *et al.* (1992) suggested a regression method, which is known as the NEGM test. The Jacobian methods have certain advantages over direct methods, such as robustness to noise and high performance in moderate-sized samples (Shintani and Linton, 2004).

The idea behind these methods is to follow two close points and to calculate the separation rate of trajectories from these points.

Suppose that x_0 and x_0' are two nearby points with distance $\|x_0 - x_0'\| = \delta_{x_0} \ll 1$. Also, suppose that δ_{x_t} is the distance between the following trajectories after T iterations. Therefore, the following statement can be made:

$$\delta_{x_t} \approx \delta_{x_0} e^{\lambda T} \tag{15}$$

In the statement above, T denotes the iteration number and λ denotes the Lyapunov exponent. Here, the λ measures the average rate of convergence or divergence of two trajectories stemming from nearby points. For small and noisy data, averaging is important.

If the system has attracting fixed points or a periodic orbit, the distance between two trajectories diminishes asymptotically with time. In an unstable system, the trajectories diverge exponentially. However, in a chaotic system, the distance between trajectories behaves erratically.

It is convenient to calculate mean exponential rate of divergence; in other words, the Lyapunov exponent is given by the following statement:

$$\lambda = \lim_{T \to \infty} \frac{1}{T} \sum_{t=1}^{T} \ln \frac{|\delta x(x_0, t)|}{|\delta x_0|} \tag{16}$$

The calculated exponents may be negative or positive, but in order to classify a system as chaotic, at least one exponent should be positive. If $\lambda \leq 0$, the system is stable. If $\lambda > 0$, the system is unstable and chaotic. The system must have at least one positive Lyapunov exponent in order to be classified as chaotic. The definition of chaotic systems involves the presence of at least one positive Lyapunov exponent. Eventually, if $\lambda = \infty$, the system is random.

3.1.4. *Surrogate data testing*

Surrogate data testing (Schreiber and Schmitz, 2000) procedure is used to test whether data are nonlinear or linear. This test is based on resampling data by using bootstrapping methods. However, certain parameters are fixed throughout the resampling process. The surrogate data testing method was presented by Theiler *et al.* (1992) and involves the following four steps:

1. The null hypothesis H_0 is created, which is connected to the mechanism that produced the observed time series.
2. An ensemble of resampled series is created that is consistent with H_0. Surrogate series are resampled series that reflect various realizations of the postulated process.
3. A discriminating statistic is calculated from surrogate series, and the distribution of this statistic is derived.
4. A significance test was used to compare the value of the discriminating statistic calculated on the original series to the distribution derived from surrogate data.

Consider the case in which we wish to check for nonlinearity in a time series. In this case, the null hypothesis H_0 is formed as follows: "The time series is created by a Gaussian linear process." Several Gaussian linear series are created as a result of this null hypothesis. From the original series and each surrogate, an appropriate statistic is calculated, and it is determined if the observed statistic differs considerably from the surrogates. If the statistics differ significantly, the null hypothesis of linearity is rejected, and it is determined that the original time series was not created by a Gaussian linear process. But the opposite of this is not correct. If there is no significant difference, we cannot say that the null hypothesis is true. Many different discriminating statistics can be utilized to distinguish original data and surrogates. Sometimes, a single discriminating statistic is not sufficient (Kugiumtzis, 2001).

To test data for various types of linear stochastic dynamics, three algorithms are proposed for generating surrogates. These are called Algorithm 1 surrogates, Algorithm 2 (AAFT) surrogates and PPS surrogates.

To test the null hypothesis that time series is driven by linear dynamics involving Gaussian white noise inputs, Algorithm 1 is used. Therefore, the model can be stated as follows:

$$x_t = \sum_{i=1}^{M} a_i x_{t-i} + \sum_{i=0}^{N} b_i \varepsilon_{t-i} \qquad (17)$$

The first sum in the above equation represents an AR model, and the second sum represents an MA model. a_i and b_i are fixed parameters (Schreiber and Schmitz, 2000). The equation above also represents linearly filtered noise (Small and Tse, 2003). However, the null hypothesis involves a large class of linear Gaussian process, not just a particular set of parameter values. To generate surrogates, Algorithm 1 utilizes the same Fourier power spectrum (Theiler, 1990; Theiler *et al.*, 1992).

If there is non-Gaussian data, to test for stochastic linear dynamics, Algorithm 2 is utilized to generate amplitude-adjusted Fourier transform (AAFT) surrogates. In this procedure, both the Fourier power spectrum and probability distribution of the data are conserved. This is obtained with static monotonic nonlinear transformation of linearly filtered noise (Small and Tse, 2003; Theiler *et al.*, 1992). Details of this method can be found in Theiler *et al.* (1992), Small and Tse (2003) and Kaplan and Glass (1995).

If the data involve aperiodic oscillations PPS surrogates can be utilized, which is proposed by Small and Tse (2003). By using PPS surrogates, the null hypothesis that aperiodic oscillations are generated by a randomly shifting periodic orbit can be tested. If this null hypothesis is rejected, a chaotic dynamic described by nonlinear nonperiodic determinism is possible. PPS surrogates retain deterministic structure such as periodic trends but remove deterministic chaotic structure.

One of the disadvantages of AAFT algorithm is that mathematical assumptions needed to carry out the static monotonic nonlinear transformation may not be possessed by real-world data. Therefore, AAFT surrogates might not comprise linear correlations in the data (Kugiumtzis, 2002) and might generate a biased Fourier power spectrum (Schreiber and Schmitz, 2000). Furthermore, with AAFT surrogates, the hypothesis of noisy trajectories could not be tested in aperiodic data. A good strategy is first, with AAFT surrogates, testing the null hypothesis that structure in the data consist of the Fourier power spectrum and the probability distribution. Thereafter, as recommended by Kugiumtzis (1999) and Kugiumtzis (2002), with other surrogates, a more general null hypothesis can be tested.

To distinguish nonlinear deterministic and linear stochastic dynamics, measures which are utilized to characterize deterministic chaos can be used. The measures used to characterize deterministic chaos include the correlation dimension, maximum Lyapunov exponent, nonlinear prediction error and entropy. These measures can be reliably used to distinguish random and deterministic chaotic structure in surrogate data testing (Kantz and Schreiber, 1997; Schreiber and Schmitz, 2000; McSharry, 2011).

In surrogate data testing, the null hypothesis that the linear stochastic dynamics can be tested with nonparametric rank-order statistics (Theiler *et al.*, 1992; Schreiber and Schmitz, 2000). In this test procedure, two parameters must be set. One of them is α, which determines the probability of false rejection, and the other one is k, which specifies the required number of surrogates. In a two-tailed hypothesis test, $S = (2k/\alpha) - 1$ surrogates are required.

For nonlinear prediction skill, an upper-tailed test is executed. In this case, the null hypothesis of linear stochastic dynamics is rejected when an attractor from the original time series predicts with more skill than its surrogate counterparts. For permutation entropy, a lower-tailed test is executed. In this case, the null hypothesis of linear stochastic dynamics is rejected when the permutation entropy calculated from the original time series is significantly less than those calculated from surrogate data. For the maximum Lyapunov exponent and correlation dimension, two-tailed tests are executed.

3.2. *Topological tools*

3.2.1. *Recurrence plots*

A recurrence plot (RP) is a square matrix which displays recurrence structure of a time series. Since recurrences are never exact in empirical time series data, a threshold parameter T is defined to evaluate recurrences. Therefore, recurrence plot is defined with the following expression:

$$RP_{ij} = \Theta(T - \|V_i(x) - V_j(x)\|) \qquad (18)$$

In the expression above, Θ represents Heaviside step function, which takes the value of one if the distance between coordinate pairs is smaller than the threshold parameter T and takes the value of zero otherwise.

Since RP is a visual tool, its interpretation involves subjective judgement of the observer. When interpreting an RP, its vertical and horizontal lines are evaluated. In an RP, if $R_{i+k,j+k} = 1$ (for $k = 1, \ldots, l$), a diagonal line appears. The length of the diagonal lines reflects predictability of the system since similar states lead to similar future. In other words, if $R_{i,j} = 1$, the probability of $R_{i+1,j+1} = 1$ is high. As a result, RPs of stochastic systems exhibit short diagonal lines or only single points and RPs of chaotic and deterministic systems exhibit longer diagonal lines. Another pattern in RPs which is interpreted is vertical (or horizontal) lines. If $R_{i,j+k} = 1$ ($R_{i+k,j} = 1$ and $k = 1, \ldots, v$), a vertical line appears. Here, v represents the length of the vertical (or horizontal) line. Vertical (or horizontal) lines emerge if the system states do not change or change very slowly. These patterns mean that states are trapped for some duration and reflect laminar states or systems are pausing at singularities. In an RP, diagonal line structures reflect chaos–order transitions and vertical (or horizontal) line structures reflect chaos–chaos transitions.

3.2.2. *Recurrence quantification analysis*

Since RPs are visual tools, their interpretation involves subjective judgement and different observers may interpret the same RP differently. To overcome this subjectivity, recurrence quantification analysis (RQA) is developed. In RQA, simple pattern recognition algorithms are applied to an RP and several measures, such as recurrence rate, determinism, laminarity entropy, trapping time and trend, are computed based on diagonal and vertical lines. The measures computed in RQA are described as follows:

DET (Determinism): Determinism is an RQA measure computed based on diagonal line patterns and reflects the predictability of a system. The term deterministic means similar current state leads to similar next state. In other words, a diagonal line is formed. Periodic systems display very long diagonal lines, chaotic systems display very short diagonal lines and stochastic systems display no diagonal lines. High value of determinism is a necessary (but not sufficient) condition for chaos. Determinism is calculated by the following formula:

$$\mathbf{DET} = \frac{\sum_{l=l_{\min}}^{N} lP(l)}{\sum_{l=1}^{N} lP(l)} \tag{19}$$

In the expression above, $P(l)$ is the frequency distribution of the lengths l of the diagonal lines.

Lmax (Length of the longest diagonal line): The length of the longest diagonal line reflects the stability of a system. More stable systems exhibit larger Lmax values. Chaotic systems exhibit low Lmax values. Lmax is inversely related to the largest positive Lyapunov exponent. Lmax can be stated as

$$\mathbf{Lmax} = \max(\{l_i; i = 1, \ldots, N_l\}) \tag{20}$$

In the expression above, N_l represents the number of diagonal lines in the RP.

Lmean (Mean length of the diagonal lines): The Lmean measure computes diagonal lines' average length. The average time that two segments of the trajectory are close to each other is revealed by this measure. This measure can be interpreted as the mean prediction time. Lmean is calculated by the following formula:

$$\mathbf{Lmean} = \frac{\sum_{l=l_{\min}}^{N} lP(l)}{\sum_{l=l_{\min}}^{N} P(l)} \tag{21}$$

LAM (Laminarity): The fraction of recurrent points producing vertical lines is measured by laminarity. Laminarity reflects laminar phases and intermittency in the system and indicates chaos–chaos transitions. Laminarity is calculated by the following formula:

$$\mathbf{LAM} = \frac{\sum_{v=v_{\min}}^{N} vP(v)}{\sum_{v=1}^{N} vP(v)} \tag{22}$$

In the expression above, $P(v)$ is the frequency distribution of the lengths v of the vertical lines.

Vmean (Trapping time): Trapping time is measured as the average length of the vertical lines. This measure reflects the average time the system is trapped in a state. Vmean is calculated by the following formula:

$$\mathbf{Vmean} = \frac{\sum_{v=v_{\min}}^{N} vP(v)}{\sum_{v=v_{\min}}^{N} P(v)} \tag{23}$$

REC (Recurrence rate): Recurrence rate is calculated as the ratio of recurrent points in the RP. It calculates the likelihood of a specific state recurring and reflects the periodicity of the system dynamics. More periodic dynamics lead to higher recurrence rates. This measure can also be used to calculate the correlation dimension. Recurrence rate is calculated by the following formula:

$$\mathbf{REC} = \frac{1}{N^2} \sum_{i,j=1}^{N} RP(i,j) \tag{24}$$

RATIO (Ratio between DET and REC): RATIO is defined as the ratio of percent determinism to percent recurrence and is quite useful in the detection of transitions between physiological states. During physiological state transitions, REC usually decreases but DET is less affected. Thus, during physiological transitions, RATIO value increases substantially but settles down again when a new quasi-steady state is achieved. RATIO is calculated as

$$\mathbf{RATIO} = N^2 \frac{\sum_{l=l_{\min}}^{N} lP(l)}{\left(\sum_{l=1}^{N} lP(l)\right)^2} \qquad (25)$$

ENTROPY (Shannon entropy of the diagonal line lengths distribution): ENTROPY measure is computed based on diagonal lines and reflects the complexity of a system. Large diversity in diagonal line lengths leads to a high ENTROPY value, and low diversity in diagonal line lengths leads to a low ENTROPY value. ENTROPY measure also expresses the amount of information required to restore the system. ENTROPY is calculated by the following formula:

$$\mathbf{ENTR} = - \sum_{l=l_{\min}}^{N} p(l) \ln p(l) \qquad (26)$$

The probability $p(l)$ that a diagonal line has exactly length l can be estimated from the frequency distribution $P(l)$ with

$$p(l) = \frac{P(l)}{\sum_{l=l_{\min}}^{N} P(l)} \qquad (27)$$

TREND (Trend of the number of recurrent points): TREND measures the degree of system stationarity. It measures the paling of the patterns of RP away from the main diagonal used for detecting drift and nonstationarity in a time series. If recurrent points are homogeneously distributed across the RP, TREND values will hover near zero. If recurrent points are heterogeneously distributed across the RP, TREND values will deviate from zero. Low values of TREND (between −5 and +5) indicate stationarity in the signal, and high values of TREND, positive or negative, indicate drift in the signal. TREND is calculated as follows:

$$\mathbf{REC}_k = \frac{1}{N-k} \sum_{\substack{j-i=k}}^{N-k} RP(i,j) \qquad (28)$$

$$\mathbf{TREND} = \frac{\sum_{i=1}^{\tilde{N}} (i - \frac{\tilde{N}}{2})(\mathrm{REC}_i - \langle \mathrm{REC}_i \rangle)}{\sum_{i=l}^{\tilde{N}} (i - \frac{\tilde{N}}{2})^2} \qquad (29)$$

In the above equation, $\langle \cdot \rangle$ means average value and $\tilde{N} < N$. $\tilde{N}$ is the maximal number of diagonals parallel to the LOI, which is considered for the calculation of TREND.

4. Dealing with Noise

Noise reduction is related to forecasting. In forecasting, it must be relied on only previous measurements. However, in noise reduction, future values can also be exploited. In noise reduction, noisy observations must be replaced by values with less noise.

In noise reduction, every observation value is decomposed to two components, namely signal and random fluctuations (noise). The classical statistical method used in noise reduction is power spectrum. The power spectrum of random noise is flat or broad, and the power spectrum of periodic or quasi-periodic signals contains sharp spectral lines. A Wiener filter can be used to separate these components.

However, for deterministic chaotic systems, this method fails because these systems have spectral properties similar to random noise. Even if the signal is matched to the spectrum, separation into noise and signal fails.

To separate signal and noise, determinism of the system can be exploited.

Suppose that the evolution of the system is deterministic defined with map f as follows:

$$x_n = f(x_{n-m}, \ldots, x_{n-1}) \tag{30}$$

However, in reality, the map f isn't known. Only noisy measurements of the signal are known as follows:

$$s_n = x_n + \eta_n \tag{31}$$

In the statement above, η_n denotes the random noise which possess fast-decaying autocorrelations and no correlations with the signal x_n. To clean a particular value in the series, prediction, $\hat{x}_n$ can be expressed as a function of previous measurements:

$$\hat{x}_n = \hat{f}(s_{n-m}, \ldots, s_{n-1}) \tag{32}$$

However, this approach will fail for chaotic systems because previous measurements are noisy and chaotic dynamics will amplify this noise. So, $\hat{x}_n$ will accumulate noise in the previous measurements. To overcome this problem, the following relation can be used to solve for middle coordination, such as $x_{n-m/2}$:

$$x_n - f(x_{n-m}, \ldots, x_{n-1}) = 0 \tag{33}$$

Since the function f isn't known and could not be solved, it is approximated by a locally constant function. To acquire an estimate $\hat{x}_{n_0-m/2}$ for the value of $x_{n_0-m/2}$, delay vectors $s_n = (s_{n-m+1}, \ldots, s_n)$ are formed and vectors which are close to s_{n_0} are determined. After that, a clean value of $\hat{x}_{n_0-m/2}$ can be chosen as the average value of $s_{n-m/2}$.

$$\hat{x}_{n_0-m/2} = \frac{1}{|\mathcal{U}_\epsilon(s_{n_0})|} \sum_{s_{n_0} \in \mathcal{U}_\epsilon(s_{n_0})} s_{n-m/2} \tag{34}$$

To measure the distances, maximum norm can be used. If the embedding dimension is large ($m > 20$) and data contain high noise level, Euclidean norm can be useful to determine the neighbors.

If the signal and noise are assumed independent, we can calculate the distances as follows:

$$(s_n - s_k)^2 = \sum_{l=0}^{m-1} (s_{n+l} - s_{k+l})^2 = \sum_{l=0}^{m-1} (x_{n+l} + \eta_{n+l} - x_{k+l} - \eta_{k+l})^2 \tag{35}$$

$$\approx (x_n - x_k)^2 + 2m\sigma^2 \tag{36}$$

As seen from above, the squared Euclidean distance between the two noisy delay vectors is equal to the squared Euclidean distance between noise-free vectors plus a constant.

According to studies in the literature, a good choice of neighborhood distance ϵ is two to three times the noise amplitude. Sometimes, the noise level can be easily determined from plot of the data. Also, it can be estimated from correlation sum. If the noise level cannot be determined with confidence, it is recommended to underestimate it. In this way, the performance can be weakened but artefacts are avoided.

In this procedure, it is convenient to choose an embedding dimension m such that both in the past and future, coordinates dynamics are deterministic. Computation burden of forming neighborhoods can be reduced by using a binary tree or a box-assisted method (Schreiber, 1995).

The term measurement noise expresses contamination of observations by errors and which is distinct from system dynamics. If a dynamical system is defined with $\mathbf{x}_{n+1} = F(\mathbf{x}_n)$, we only measure values as $s_n = s(\mathbf{x}_n) + \eta_n$. Here, $s(\mathbf{x}_n)$ denotes a smooth function which matches points in the phase space with scalars and η_n denotes random noise. Thus, the series $\{\eta\}_n$ is called as measurement noise.

On the other hand, in the dynamical noise, errors are given as input to the function determining system dynamics as described below:

$$\mathbf{x}_{n+1} = F(\mathbf{x}_n + \eta_n) \qquad (37)$$

Discrimination of measurement and dynamical noise *a posteriori* is difficult by using only data. In chaotical systems, the dynamical and measurement noise can be mapped onto each other (Grebogi *et al.*, 1990). The nature of noise determines the predictability and other properties of the data. Adding a noise into the function $\mathbf{F}$ can be recovered by changing parameters of $\mathbf{F}$ slightly. On the other hand, if parameters of $\mathbf{F}$ fluctuate stochastically, parameters can be compensated by noise. In the situation of dynamical noise, all system parameters are prone to random perturbations. In practice, dynamical noise creates much greater problems than measurement noise.

Noise in the data adversely effects performance of prediction in deterministic systems. Also, noise leads to deterioration of self-similarity of the system's strange attractor. In the presence of noise, power-law scaling of correlation sum $(C(\epsilon))$ is broken. In the case of Gaussian uncorrelated noise, the maximum acceptable noise degree for estimation of the correlation dimension is around 2%. The situation is not better for entropies and Lyapunov exponents (Schreiber and Kantz, 1996). There is a tradeoff between invariance properties of these measures and the noise. Moreover, embedding theorems are only valid for data with infinite resolution. Tolerable noise level depends on the detail of underlying system and the measurement. If a result sensitively depends on the details of the embedding procedure, it is recommended to carefully interpret the results.

There exist many beneficial measures which are not invariant, such as mean, variance, autocorrelation function and power spectrum. Non-invariant quantities can be carefully used as relative measures.

Some effects of noise can be enumerated as follows (Kantz and Schreiber, 1997):

- Self-similarity is visibly broken.
- Nearby trajectories diverge diffusively rather than exponentially.
- A phase space reconstruction appears as high dimensional on small length scales.
- The prediction error is found to be bounded from below no matter which prediction method is used and to how many digits the data are recorded.

Nonlinear noise reduction and nonlinear prediction are closely related. Suppose that data generated by a deterministic system but are measured with some noise η. We can state that

$$s_n = x_n + \eta_n, \quad x_{n+1} = F(x_{n-m+1}, \ldots, x_n) \qquad (38)$$

To predict a value and reduce noise, equation F which defines the deterministic dynamics, must be estimated. To predict a future value, such as $\hat{s}_{N+1}$, the last data point s_N is fed to the estimated function $\hat{F}$ and future value is calculated. To obtain a cleaned data series, such as $\hat{x}_n$, $n = 1, \ldots, N$, the dynamical structure is exploited.

In the literature, several noise-reduction methods are proposed. These methods differ in terms of approximation to the dynamics, adjustment of the trajectory and linkage of the adjustment and approximation steps. Criteria for selecting a noise reduction algorithm are robustness, ease of application and computer time and memory requirement.

For now, assume that a dynamical equation of motion F is known or a reliable estimate of $\hat{F}$ is available and dimension is one. If we try to model noise as $\hat{x}_{n+1} = \hat{F}(s_n)$, we could not reduce noise. By evaluating the following statement

$$\hat{x}_{n+1} = \hat{F}(\hat{x}_{n-m+1}, \ldots, \hat{x}_n) + \eta_n' \qquad (39)$$

a measure for error η_n' can be constituted in a cleaned trajectory. Therefore, we could numerically solve the equation above for $\hat{x}_n$ by setting all η'_n to zero. However, since equations for chaotic maps are ill-conditioned, it is hard to solve this numerical problem numerically. Even so, if exact dynamics are known, it is possible to avoid some of the difficulties (Hammel, 1990; Farmer and Sidorowich, 1991).

A more convenient method is to minimize mean square error instead of setting all the η'_n to zero:

$$e^2 = \sum_{n=1}^{N-1} \eta'^2_n = \sum_{n=1}^{N-1} (\hat{x}_{n+1} - \hat{F}(\hat{x}_n))^2 \tag{40}$$

This nonlinear minimization problem can be solved numerically by gradient descent:

$$\hat{x}_n = s_n - \frac{\alpha}{2} \frac{\partial e^2}{\partial s_n} \tag{41}$$

$$= (1 - \alpha)s_n + \alpha[\hat{F}(s_{n-1}) + \hat{F}'(s_n)(s_{n+1} - \hat{F}(s_n))] \tag{42}$$

In the equations above, α denotes step size. This intuitive method is presented by Davies (1992) and consists of two steps. In the first step, the dynamics are approximated, and in the second step, the trajectory is adjusted to dynamics. In the second step, to estimate $\hat{x}_n$, both the past and future information is used. When approximate dynamics and derivatives are available, this method is useful. However, noise hinders the construction of accurate global models. When enough data are present for local approximations, the locally linear approach shows good performance. When $\hat{F}$ is estimated as locally linear maps, the mean square error function can be minimized by solving a linear equation system. This procedure consists of multiple steps and in each step, we go to the right direction and fit a better map for $\hat{F}$ before the next step (Kostelich and Yorke, 1988).

Usage of local linear maps is the classical method to approximate the equations for a deterministic dynamical system. To accomplish this, the only assumption we must make is that map $\hat{\mathbf{F}}$ is a smooth function and at least piecewise differentiable. In this way, $\hat{\mathbf{F}}$ can be linearized in the vicinity of $\mathbf{x}_n$ as follows:

$$\hat{F}(\mathbf{x}) = \hat{\mathbf{F}}(\mathbf{x}_n) + \mathbf{J}_0(\mathbf{x} - \mathbf{x}_n) + O(\|\mathbf{x} - \mathbf{x}_n\|^2) \tag{43}$$

$$\approx \mathbf{J}_n\mathbf{x} + \mathbf{b}_n \tag{44}$$

In the statement above, $\mathbf{J}_n$ denotes the Jacobian matrix of $\hat{\mathbf{F}}$ in $\mathbf{x}_n$. It is shown by Eckmann and Ruelle (1985) that the above linear approximation can be obtained from least squares fit of the time series. This linear approximation is utilized by Eckmann *et al.* (1986) to estimate Lyapunov exponents. Also, this approximation is used for prediction by Farmer and Sidorowich (1987).

If the description of dynamics is available in an explicit form, nonlinear noise reduction can be realized with a separate trajectory adjustment step. Since chaotic systems are sensitive to initial conditions, this is necessary. To avoid this instability, a new formulation can be developed. In this formulation, by solving a minimization problem, each delay vector is individually corrected to be compatible with the dynamics. In this way, by avoiding corrections to the earliest and latest coordinates of each single delay vector, stability is reached. By determining an appropriate metric in phase space, such dangerous corrections are avoided. A local projective correction scheme is described in the following.

An $(m-1)$-dimensional map $x_n = F(x_{n-m+1},\ldots,x_{n-1})$ can be written as $\tilde{F}(x_{n-m+1},\ldots,x_n) \equiv \tilde{F}(\mathbf{x}_n) = 0$ and can be linearized as

$$\mathbf{a}^{(n)} \cdot \mathbf{R}(\mathbf{x}_n - \bar{x}^{(n)}) = 0 + O(\|\mathbf{x}_n - \bar{x}^{(n)}\|^2) \tag{45}$$

In the statement above, $\bar{\mathbf{x}}^{(n)} = |\mathcal{U}_n|^{-1}\sum_{n'\in\mathcal{U}_n}\mathbf{x}_{n'}$ denotes delay vectors' center of mass in a neighborhood of $\mathbf{x}_n$, $\mathbf{R}$ is the diagonal weight matrix and $\mathbf{a}^{(n)}$ denotes the direction in phase space. With the help of the diagonal weight matrix, noise reduction on the most stable middle coordinates can be achieved. This can be realized by setting R_{11} and R_{mm} a large number and for all other diagonals, $R_{ii} = 1$. Orthogonal projections (Cawley and Hsu, 1992; Sauer, 1992) are obtained when $\mathbf{R} = \mathbf{1}$. If the series s_n contains noise, this relationship will not exactly hold but with some error:

$$\mathbf{a}^{(n)} \cdot \mathbf{R}(\mathbf{s}_n - \bar{s}^{(n)}) = \eta_n \tag{46}$$

In the following, the notation (n) will be suppressed. The linear equation above is valid only locally, and the direction $\mathbf{a}^{(n)}$ and center of mass $\bar{\mathbf{s}}^{(n)}$ depend on the position in phase space.

In practice, the correct dimension $m-1$ usually is not known. Furthermore, the formation of delay coordinates to construct dynamics deterministically increases the minimum value of m. In general, we form delay coordinates in m dimensions, but the dynamics have a lower-dimensional manifold whose dimension is m_0. Therefore, up to

$Q = m - m_0$ mutually independent sub-spaces $\mathbf{a}^q$, $(q = 1, \dots, Q)$ can be found which satisfies the equations above. Linear space spanned by these Q vectors are called nullspace at point $\mathbf{x}_n$. Since the noise-free attractor does not stretch out to this space, the part of $\mathbf{s}_n$ that we find in it must be because of noise. The locally projective noise reduction algorithm identifies this null space and afterward eliminates the relating part of $\mathbf{s}_n$.

In the nonlinear noise reduction, it is important to evaluate whether noise has been really reduced. Assessment of noise-reduction level depends on assumptions about the nature of the noise. Local projective noise-reduction algorithm is efficient when the data consist of deterministic part plus random noise. Therefore, the effectiveness of the noise-reduction algorithm can be investigated by testing whether data become more deterministic and obtained noise is random.

To use in algorithms, two error measures can be utilized. If the noise-free data are known, the error can be calculated by root mean square distance between measured signal s_n and noise-free data s_n^0 as follows:

$$e_0 = \sqrt{\frac{1}{N} \sum_{n=1}^{N} (s_n - s_n^0)^2} \tag{47}$$

The root mean square distance $\hat{e}_0$ can be also calculated for the cleaned data $\hat{x}_n$. If $\hat{e}_0$ is less than e_0, we can conclude that the noise has been reduced. Another error measure can be calculated if the equation defining the evolution of dynamical system is known, such as $\mathbf{x}_n^0 = \mathbf{F}(\mathbf{x}_{n-1}^0)$. This measure quantifies the deviation from deterministic behavior and calculated as

$$e_{\mathrm{dyn}} = \sqrt{\frac{1}{N} \sum_{n=1}^{N} (\mathbf{x}_n - \mathbf{f}(\mathbf{x}_{n-1}))^2} \tag{48}$$

This dynamical error can be calculated before and after execution of the noise-reduction algorithm and the results can be compared.

The above measures are beneficial if the true dynamics are known. If a parameter in the dynamical equations is slightly changed, the output of a noise-reduction algorithm might be much cleaner. Also, a noise-reduction algorithm can shift the data in a systematic way. In this case, larger values for e_0 and e_{dyn} are obtained. If the coordinates are changed, data might become less noisy. Computer-generated data might lead to these misinterpretations. It is assumed that every dynamical system has original equations of motion, and if these equations are known, the system can be described and understood perfectly. In computer-generated data, the equations defining the system are known. However, in reality, there exist many ways to generate same data.

For data obtained from experiments, the exact dynamical equations and noise-free signal are not known. So, measures such as e_0 and e_{dyn} are not applicable. Therefore, algorithms and measures based on only measured signal are needed. The noise level can be determined by using the correlation integral. With this procedure, amplitude of noise can be estimated before and after noise reduction and successful cleaning can be verified.

If a noise-reduction algorithm effectively removes the noise, then the cleaned data must possess better short-term predictability than the uncleaned data. This difference can be assessed by calculating out-of-sample prediction error before and after the noise reduction. In this case, data are split into two parts for fitting a nonlinear mapping $\hat{\mathbf{F}}(\mathbf{s})$ and the prediction error can be calculated for N_p data points:

$$e = \sqrt{\frac{1}{N_p} \sum_{n=N-N_p+1}^{N} \left(\mathbf{s}_n - \hat{\mathbf{F}}(\mathbf{s}_{n-1})\right)^2} \qquad (49)$$

It is important to obtain out-of-sample error because zero in-sample prediction error can be always obtained by interpolation. If the data is too short, take-one-out statistics can be utilized to acquire out-of-sample error (Grassberger *et al.*, 1991; Efron, 1982).

In low-dimensional chaotic system, one-step prediction error shows up because of three effects:

1. Measured value s_n contains noise in an amount of noise level.
2. Arguments pertaining to prediction function contain noise. This results in an error proportional to the noise level. However, this noise is exaggerated because of sensitivity to the initial conditions.

3. The fitted prediction function might not be accurate. This accuracy is influenced by noise level.

The first and second effects are proportional to the noise level, but the magnitude of the third effect is generally unknown. It is expected that the deviation from true dynamics scale up monotonically with the noise level. If all three effects are present, e scales up faster than linearly with the noise level. When e before noise reduction and $\hat{e}$ after noise reduction are compared, an upper bound on the level of noise reduction can be obtained. When $\hat{e} < e$, the described data are more self-consistent.

Chaotic data are described by broadband power spectra. So, it is hard to clean noise by utilizing procedures based on power spectra. However, in some cases, it is possible to distinguish signal and noise in some part of the spectrum, for instance, around a dominant frequency or at high frequencies. Success of nonlinear noise-reduction methods should be apparent in these parts of the power spectrum. For noise-reduction, the usage of the power spectrum is not recommended because it is a linear method and insensitive to broadband spectra.

If the measured data are the sum of deterministic signal and noise, then the corrections to clean data should imitate a random process. Therefore, it is convenient to investigate the distribution of the corrections. It is expected that corrections follow a normal distribution. Furthermore, cross-correlations between the signal and corrections can be investigated. If the noise-reduction procedure is successful, cross-correlations should be small.

Signal processing methods can be used to filter noise from the data and unmask drivers of complex behavior. One of these methods is singular spectrum analysis (SSA). SSA is a data-driven method that can detect drivers of structural variation in noisy time series without any theoretical presumption about the source of irregularity (Elsner and Tsonsis, 2013; Golyandina *et al.*, 2001; Golyandina *et al.*, 2018; Hassani and Mahmoudvand, 2018; Hassani and Zhigljavsky, 2009; Hassani and Thomakos, 2010). In SSA, signal (structured variation), such as trend and oscillatory components, and noise (unstructured variation) are separated. This separation takes place in three steps: matrix decomposition, grouping and time series reconstruction. These steps can be realized with two different methods, namely, singular value decomposition (SVD)-SSA and Toeplitz-SSA. Each method utilizes a different method for matrix decomposition. Besides filtering noise, SSA has additional benefits, such as easing nonstationarity and removing slow-moving trend. Another application of SSA is to fill in missing observations in the data. SSA can fill in missing observations by utilizing structural variation in the whole time series rather than only looking at surrounding observations.

5. Forecasting in Nonlinear and Chaotic Time Series

When modeling a system, the aim is usually to establish the equations describing the system. When these equations are derived from measurements, the obtained model is called phenomenological model. However, in other cases, the dynamical model is derived from first principles.

To set up a useful phenomenological model, specialized knowledge about the system is required. On the other hand, some models can be constructed using only time series data. In this way, dynamical laws governing a system can be obtained from a sequence of observations. However, due to incompleteness of the data, the solution of this problem may not be unique. Under these circumstances, the simplest model consistent with the observations is preferred.

A successful model should reproduce the data in a statistical manner. The first step in this approach is determining a model equation including free parameters, and the second step is fitting to observed dynamics.

A correct time series model can be used for prediction, noise reduction, control and generation of artificial data. Construction of a correct model and construction of the optimal predictor may eventuate in different mathematical expressions. Some predictive methods may not yield a globally valid model.

Modeling is an art which differs from computation of a measure. So, there is no strict algorithm for model building.

For time series, there can be different sources of predictability. One example is linear correlations which are revealed by autocorrelation function. In this case, an AR, MA or a combination of these two can be chosen as a linear stochastic model.

Another source of predictability is determinism. In this case, equal initial states will evolve to an equal future and similar initial states will evolve to similar future states, at least over short times. For these kinds of dynamics, a deterministic model must be constructed. An irregular time series can be explained by a deterministic chaotic model. Chaotic signals exhibit autocorrelation functions, which decay exponentially fast, and a broadband component in spectra. For this reason, autocorrelations cannot be exploited for construction of a deterministic model.

In TSA, forecasting is an important problem. If some pattern is found in a time series, it can be used to improve prediction. A theory can be justified by successful prediction. If a time series is steady and does not fluctuate, it is easy to predict. For this kind of time series, the last observation is a proper forecast for the next observation. For changing signals, this can be still a convenient forecast. These kinds of signals are called persistent. Forecasting periodically changing systems is also easy. Forecasting independent random numbers are easy as well. In this case, the best prediction is the mean value. Some signals fall between these two situations. Although they are not periodic, they contain some structure which can be used for better predictions.

There are some error measures to quantify the success of predictions. One of them is root mean square (RMS) prediction error. If the predicted values are $\hat{s}_n$ and the actual values are s_n, then RMS prediction error is computed as follows:

$$e = \sqrt{\langle (\hat{s}_n - s_n)^2 \rangle} \tag{50}$$

In the statement above, $\langle \cdot \rangle$ means average. This error is minimized for independent random numbers if prediction is taken as average. Other examples of error measures are mean absolute error $\langle |\hat{s}_n - s_n|$ and logarithmic error $\langle \log |\hat{s}_n - s_n| \rangle$.

The prediction accuracy can be increased if statistical distribution of the values are known. The mean value and probability distributions do not take into account the correlations in time. If correlations in time are present in the series, they can be used to improve predictions.

Linear correlations in time are the most usual sources of predictability. If strong correlations are present in the series, the next observation can be predicted by a linear combination of previous observations.

In nonlinear deterministic dynamical systems, a different type of temporal correlation exists. Dynamical systems can be defined by discrete time maps or differential equations:

$$\mathbf{x}_{n+1} = \mathbf{F}(\mathbf{x}_n), \frac{d}{dt}\mathbf{x}(t) = \mathbf{f}(\mathbf{x}(t)) \tag{51}$$

In both forms, the future states are perfectly determined by the present state. However, in the cases of chaotic systems, uncertainty of present state is amplified during the evolution of the system even if deterministic evolution law is known. Therefore, forecasting of far future is impossible in chaotic systems. However, short-term forecasts can be feasible in chaotic systems. Nonlinear deterministic dynamics can be evaluated by nonlinear statistics. In this context, nonlinear correlations can be exploited for predictions.

A simple prediction algorithm that utilizes deterministic feature of the signal can be developed (Kantz and Schreiber, 1997). For this algorithm, it is assumed that data are generated by a dynamical system.

If discrete time map $\mathbf{F}$ of the dynamical system is known, it can be used for prediction. However, in reality, usually $\mathbf{F}$ is not known. Therefore, some assumptions should be made about its properties. If it is assumed that $\mathbf{F}$ is continuous, a simple prediction algorithm can be constituted. To predict the future value $\mathbf{x}_{N+1}$, given the present value $\mathbf{x}_N$, a list of all past values $\mathbf{x}_n$ with $n < N$, which are closest to $\mathbf{x}_N$ with respect to some norm, are searched. If the value at time n_0 is similar to the present value (near in the phase space), continuity of $\mathbf{F}$ implies that $\mathbf{x}_{n_0+1}$ should also be near to $\mathbf{x}_{N+1}$. If the system is evaluated for a long time, there should be values in the past which are close to present value. Therefore, the predicted value $\hat{\mathbf{x}}_{N+1} = \mathbf{x}_{n_0+1}$ should be close the true value. This approach is called "Lorenz's method of analogues" because it was suggested by Lorenz (1969).

However, in reality, actual values $\mathbf{x}_N$ are not usually measured. Quantities, which are functions of these values, are observed. Usually, scalar measurements are present, such as

$$s_n = s(\mathbf{x}_n), \quad n = 1, \ldots, N \tag{52}$$

In this case, the measurement function s, like $\mathbf{F}$, is also unknown. Evidently, the measurement function s cannot be inverted but vectors equivalent to the original values can be acquired by utilizing delay reconstruction as follows:

$$\mathbf{s}_n = \left(s_{n-(m-1)\tau}, s_{n-(m-2)\tau}, \ldots, s_{n-\tau}, s_n\right) \tag{53}$$

In the statement above, τ denotes the delay time and m denotes the embedding dimension. In this way, for measurements $s_1, \ldots, s_N$, corresponding delay vectors $\mathbf{s}_{(m-1)\tau+1}, \ldots, \mathbf{s}_N$ can be constructed. In this method, to predict a future measurement $s_{N+\Delta n}$, embedding vector $\mathbf{s}_{n_0}$ closest to $\mathbf{s}_N$ is determined and $s_{n_0+\Delta n}$ is selected as a predictor. Kennel and Isabelle (1992) used this method to test for determinism.

However, in reality, every measurement has some finite resolution and stored in a discretized form. If we call the typical size of finite resolution as σ, then the distances between points will contain an uncertainty of the size of σ. Hence, all points within a radius σ should be taken as equally good predictors. In this case, arithmetic mean of the individual predictions can be used. So, the resolution of the measurements should be considered as another parameter for the prediction method.

Eventually, the prediction algorithm can be described as follows. First, for a scalar time series $s_1, \ldots, s_N$, an embedding dimension m and a delay time τ are determined and delay vectors are formed, such as $\mathbf{s}_{(m-1)\tau+1}, \ldots, \mathbf{s}_N$. To predict a value time Δn ahead of N, we choose a parameter ϵ and form a neighborhood $\mathcal{U}_\epsilon(\mathbf{s}_N)$ of radius ϵ around the point $\mathbf{s}_N$. Then, the prediction is obtained as the average of individual predictions:

$$\hat{s}_{N+\Delta n} = \frac{1}{|\mathcal{U}_\epsilon(\mathbf{s}_N)|} \sum_{\mathbf{s}_n \in \mathcal{U}_\epsilon(\mathbf{s}_N)} s_n + \Delta n \tag{54}$$

In the statement above, $|\mathcal{U}_\epsilon(\mathbf{s}_N)|$ denotes the number of elements in the neighborhood $\mathcal{U}_\epsilon(\mathbf{s}_N)$.

If a prediction is true, it may be just a coincidence. In order to justify the significance of predictions, errors of many predictions should be evaluated.

There is a distinction between in-sample prediction error and out-of-sample prediction error. To evaluate the prediction performance, the out-of-sample prediction error should be considered. To compute these prediction errors, the data should be divided into a training set and a test set. The parameters of the prediction algorithm are optimized on the training set and performance is measured on the test set. In practice, out-of-sample errors are usually much larger than in-sample errors.

6.   Conclusion

In this study, we discussed nonlinear and chaotic TSA methods. Chaos is a nonlinear deterministic process which resembles a random stochastic process. Chaotic dynamics must be necessarily a result of a nonlinear model. A linear model can only produce four types of behavior: oscillatory and stable, oscillatory and explosive, non-oscillatory and stable, and non-oscillatory and explosive. To classify a continuous system as chaotic, it must have at least three dimensions. Discrete systems can exhibit chaotic dynamics even with single dimension. There are many methods to test whether a time series is generated from a chaotic system. These methods are divided into two subclasses, namely metric and topological tools. Metric tools provide some statistics or numeric values which are used to judge whether a time series is chaotic or not. Metric tools include BDS test, correlation dimension, Lyapunov exponents and surrogate data test. Topological tools are based on recurrence plots. Recurrence plots can be analyzed visually or by using recurrence quantification analysis. The main factor that complicates TSA is noise in the data. In this work, we presented several ways to clean noise in the data. The challenging task in the chaotic TSA is forecasting. Since chaotic systems exhibit sensitivity to initial conditions, forecasting is not possible in the long term. In this work, we presented a forecasting method for short-term predictions.

References

Bollerslev, T. (1986). Generalized autoregressive conditional heteroskedasticity. *Journal of Econometrics*, **31**(3): 307–327.

Brock, W.A. (1986). Distinguishing random and deterministic systems: Abridged version. *Journal of Economic Theory*, **40**(1): 168–195.

Brock, W.A., Hsieh, D.A., LeBaron, B.D. and Brock, W.E. (1991). *Nonlinear Dynamics, Chaos, and Instability: Statistical Theory and Economic Evidence*, (MIT Press).

Brock, W.A., Scheinkman, J.A., Dechert, W.D. and LeBaron, B. (1996). A test for independence based on the correlation dimension. *Econometric Reviews*, **15**(3): 197–235.

Buzug, T. and Pfister, G. (1992). Comparison of algorithms calculating optimal embedding parameters for delay time coordinates. *Physica D: Nonlinear Phenomena*, **58**(1–4): 127–137.

Casdagli, M., Eubank, S., Farmer, J.D. and Gibson, J. (1991). State space reconstruction in the presence of noise. *Physica D: Nonlinear Phenomena*, **51**(1–3): 52–98.

Cawley, R. and Hsu, G.H. (1992). Local-geometric-projection method for noise reduction in chaotic maps and flows. *Physical Review A*, **46**(6): 3057.

Davies, M. (1993). Noise reduction by gradient descent. *International Journal of Bifurcation and Chaos*, **3**(01): 113–118.

Devaney, R. (1989). *An Introduction to Chaotic Dynamical Systems*, (Addison-Wesley).

Eckmann, J.P. and Ruelle, D. (1985). Ergodic theory of chaos and strange attractors. In B. R. Hunt, J. A. Kennedy, and H. E. Nusse (Eds.): *The Theory of Chaotic Attractors*, (pp. 273–312) (Springer).

Efron, B. (1982). *The Jackknife, the Bootstrap and Other Resampling Plans*, (SIAM).

Elsner, J.B. and Tsonis, A.A. (2013). *Singular Spectrum Analysis: A New Tool in Time Series Analysis*, (Springer).

Engle, R.F. (1982). Autoregressive conditional heteroskedasticity with estimates of the variance of UK inflation. *Econometrica.*, **50**: 987–1007.

Farmer, J.D and Sidorowich, J.J. (1991). Optimal shadowing and noise reduction. *Physica D: Nonlinear Phenomena*, **47**(3): 373–392.

Fraser, A.M. and Swinney, H.L. (1986). Independent coordinates for strange attractors from mutual information. *Physical Review A*, **33**(2): 1134–1140.

Gibson, J.F., Farmer, J.D., Casdagli, M. and Eubank, S. (1992). An analytic approach to practical state space reconstruction. *Physica D: Nonlinear Phenomena*, **57**(1–2): 1–30.

Gilmore, C.G. (1993a). A new test for chaos. *Journal of Economic Behavior & Organization*, **22**(2): 209–237.

Gilmore, C.G. (1993b). A new approach to testing for chaos, with applications in finance and economics. *International Journal of Bifurcation and Chaos*, **3**(03): 583–587.

Golyandina, N., Nekrutkin, V. and Zhigljavsky, A. (2001). *Analysis of Time Series Structure: SSA and Related Techniques*, (CRC Press).

Golyandina, N., Korobeynikov, A. and Zhigljavsky, A. (2018). *Singular Spectrum Analysis with R*, (Springer).

Grassberger, P. and Procaccia, I. (1983). Characterization of strange attractors. *Physical Review Letters*, **50**(5): 346.

Grassberger, P., Schreiber, T. and Schaffrath, C. (1991). Nonlinear time sequence analysis. *International Journal of Bifurcation and Chaos*, **1**(03): 521–547.

Grebogi, C., Hammel, S.M., Yorke, J.A. and Sauer, T. (1990). Shadowing of physical trajectories in chaotic dynamics: Containment and refinement. *Physical Review Letters*, **65**(12): 1527.

Hammel, S.M. (1990). A noise reduction method for chaotic systems. *Physics Letters A*, **148**(8–9): 421–428.

Hassani, H. and Mahmoudvand, R. (2018). *Singular Spectrum Analysis: Using R*, (Springer).

Hassani, H. and Thomakos, D. (2010). A review on singular spectrum analysis for economic and financial time series. *Statistics and Its Interface*, **3**(3): 377–397.

Hassani, H. and Zhigljavsky, A. (2009). Singular spectrum analysis: Methodology and application to economics data. *Journal of Systems Science and Complexity*, **22**(3): 372–394.

Hsieh, D.A. and LeBaron, B. (1988). *Finite Sample Properties of the BDS Statistic. Working Paper*, (University of Chicago and University of Wisconsin).

Kantz, H. and Schreiber, T. (1997). *Nonlinear Time Series Analysis*, (Cambridge University Press).

Kaplan, D. and Glass, L. (1995). *Understanding Nonlinear Dynamics* (Springer).

Kennel, M.B. and Isabelle, S. (1992). Method to distinguish possible chaos from colored noise and to determine embedding parameters. *Physical Review A*, **46**(6): 3111.

Kennel, M.B., Brown, R. and Abarbanel, H.D. (1992). Determining embedding dimension for phase-space reconstruction using a geometrical construction. *Physical Review A*, **45**(6): 3403–3411.

Kostelich, E.J. and Yorke, J.A. (1988). Noise reduction in dynamical systems. *Physical Review A*, **38**(3): 1649.

Kugiumtzis, D. (1997). Assessing different norms in nonlinear analysis of noisy time series. *Physica D: Nonlinear Phenomena*, **105**(1–3): 62–78.

Kugiumtzis, D. (1999). Test your surrogate data before you test for nonlinearity. *Physical Review E*, **60**(3): 2808.

Kugiumtzis, D. (2001). On the reliability of the surrogate data test for nonlinearity in the analysis of noisy time series. *International Journal of Bifurcation and Chaos*, **11**(07): 1881–1896.

Kugiumtzis, D. (2002). Surrogate data test on time series. In A. Soofi and L. Cao (Eds.): *Modelling and Forecasting Financial Data, Techniques of Nonlinear Dynamics*, (pp. 267–282) (Springer).

Liebert, W. and Schuster, H.G. (1989). Proper choice of the time delay for the analysis of chaotic time series. *Physics Letters A*, **142**(2–3): 107–111.

Mane, R. (1981). On the dimension of the compact invariant set of certain non-linear maps. In D. A. Rand and L. S. Young (Eds.): *Dynamical Systems and Turbulence: Proceedings of a Symposium Held at the University of Warwick 1979/80*, (pp. 230–242) (Springer).

McCaffrey, D.F., Ellner, S., Gallant, A.R. and Nychka, D.W. (1992). Estimating the Lyapunov exponent of a chaotic system with nonparametric regression. *Journal of the American Statistical Association*, **87**(419): 682–695.

McSharry, P. (2011). The danger of wishing for chaos. In S. Guastello and R. Gregson (Eds.): *Nonlinear Dynamical Systems Analysis for the Behavioral Sciences Using Real Data*, (pp. 539–558) (CRC Press).

Packard, N.H., Crutchfield, J.P., Farmer, J.D. and Shaw, R.S. (1980). Geometry from a time series. *Physical Review Letters*, **45**(8): 712.

Provenzale, A., Smith, L.A., Vio, R. and Murante, G. (1992). Distinguishing between low-dimensional dynamics and randomness in measured time series. *Physica D: Nonlinear Phenomena*, **58**(1–4): 31–49.

Rosenstein, M.T., Collins, J.J. and De Luca, C.J. (1994). Reconstruction expansion as a geometry-based framework for choosing proper delay times. *Physica D: Nonlinear Phenomena*, **73**(1): 82–98.

Sauer, T. (1992). A noise reduction method for signals from nonlinear systems. *Physica D: Nonlinear Phenomena*, **58**(1–4): 193–201.

Sauer, T., Yorke, J.A. and Casdagli, M. (1991). Embedology. *Journal of Statistical Physics*, **65**(3): 579–616.

Schreiber, T. (1995). Efficient neighbor searching in nonlinear time series analysis. *International Journal of Bifurcation and Chaos*, **5**(02): 349–358.

Schreiber, T. and Kantz, H. (1996). Observing and predicting chaotic signals: Is 2% noise too much? In Y. A. Kravtsov and J. B. Kadtke (Eds.): *Predictability of Complex Dynamical Systems*, (pp. 43–65) (Springer).

Schreiber, T. and Schmitz, A. (2000). Surrogate time series. *Physica D: Nonlinear Phenomena*, **142**(3–4): 346–382.

Shintani, M. and Linton, O. (2004). Nonparametric neural network estimation of Lyapunov exponents and a direct test for chaos. *Journal of Econometrics*, **120**(1): 1–33.

Small, M. and Tse, C.K. (2003). Detecting determinism in time series: The method of surrogate data. *IEEE Transactions on Circuits and Systems I: Fundamental Theory and Applications*, **50**(5): 663–672.

Takens, F. (1981). Detecting strange attractors in turbulence. In D.A. Rand and L.S. Young (Eds.): *Dynamical Systems and Turbulence: Proceedings of a Symposium Held at the University of Warwick 1979/80*, (pp. 366–381) (Springer).

Theiler, J. (1986). Spurious dimension from correlation algorithms applied to limited time-series data. *Physical Review A*, **34**(3): 2427–2432.

Theiler, J. (1990). Estimating the fractal dimension of chaotic time series. *Lincoln Laboratory Journal*, **3**: 63–86.

Theiler, J., Eubank, S., Longtin, A., Galdrikian, B. and Farmer, J.D. (1992). Testing for nonlinearity in time series: The method of surrogate data. *Physica D: Nonlinear Phenomena*, **58**(1–4): 77–94.

Whitney, H. (1936). Differentiable manifolds. *Annals of Mathematics*, **37**: 645–680.

Wolf, A., Swift, J.B., Swinney, H.L. and Vastano, J.A. (1985). Determining Lyapunov exponents from a time series. *Physica D: Nonlinear Phenomena*, **16**(3): 285–317.

Chapter 5

A Fiducial-based Test for the Equality of Location Parameters

Gamze Güven[*,†,§], **Özge Gürer**[‡,¶], **Hatice Samkar**[†,‖],
and **Birdal Şenoğlu**[‡,**]

†*Department of Statistics, Eskisehir Osmangazi University,*
Eskisehir, Turkey

‡*Department of Statistics, Ankara University,*
Ankara, Turkey

§*gamzeguven@ogu.edu.tr*
¶*otanju@ankara.edu.tr*
‖*hfidan@ogu.edu.tr*
***senoglu@science.ankara.edu.tr*

Motivated by Fisher's fiducial inference, a new test is proposed for testing the equality of location parameters of Weibull populations when the scale parameters are unequal. It is defined based on the sum of the squared differences between the estimates of location parameters and weighted mean of the estimates of location parameters. In estimating the model parameters, modified maximum likelihood (MML) methodology proposed by Tiku is used. Resulting estimators are asymptotically equivalent to the maximum likelihood (ML) estimators; therefore, they have all the asymptotic properties of the ML estimators, such as unbiasedness, efficiency and consistency. The proposed test is constructed using MML estimators and a pivotal quantity. A comprehensive Monte

---

[*]Corresponding author.

Carlo simulation study is conducted to compare this test with the corresponding test based on the well-known least squares (LS) estimators with respect to the power criterion. Simulation results show that the proposed test outperforms the LS-based test in terms of the power criterion.

1.   Introduction

Weibull distribution is one of the well-known skew distributions in the literature. It is widely used in nearly all scientific disciplines, such as engineering, statistics, geophysics, medicine, finance, insurance and biology.

Probability density function (PDF) and cumulative distribution function (CDF) of Weibull distribution are defined as

$$f(y; \mu, \sigma, p) = \frac{p}{\sigma} \left(\frac{y - \mu}{\sigma} \right)^{p-1} \exp \left\{ - \left(\frac{y - \mu}{\sigma} \right)^p \right\},$$

$$y > \mu, \quad \sigma > 0, \quad p > 0 \tag{1}$$

and

$$F(y; \mu, \sigma, p) = 1 - \exp \left\{ - \left(\frac{y - \mu}{\sigma} \right)^p \right\} \tag{2}$$

respectively. Here, μ is the location, σ is the scale and p is the shape parameter. The hazard rate function corresponding to Eq. (2) is given as follows:

$$H(y; \mu, \sigma, p) = \frac{p}{\sigma} \left(\frac{y - \mu}{\sigma} \right)^{p-1} \tag{3}$$

In this study, we consider the problem of testing the hypothesis

$$H_0: \mu_1 = \mu_2 = \cdots = \mu_a \quad \text{vs.} \quad H_1: \mu_i \neq \mu_j \quad \exists i \neq j \tag{4}$$

where a is the number of Weibull populations. The reason for using Weibull distribution in the current study is its flexibility for modeling data having a broad range of asymmetric distributions. If $p = 1$ and $p = 2$, it reduces to the exponential and Rayleigh distributions, respectively. Furthermore, Weibull distribution is reverse J-shaped

Table 1. Skewness and kurtosis values of Weibull distribution for various values of the shape parameter p.

p	1.5	2	2.5	3	4	6
$\sqrt{\beta_1}$	1.064	0.631	0.358	0.168	-0.087	-0.373
β_2	4.365	3.246	2.858	2.705	2.752	3.035

and hazard rate function is decreasing if p is less than one, whereas it is bell-shaped and hazard rate function is increasing if p is greater than one, Acıtaş (2019); Cohen and Whitten (1982); Lai *et al.* (2006). Skewness and kurtosis values (i.e. $\sqrt{\beta_1} = \mu_3/\mu_2^{3/2}$ and $\beta_2 = \mu_4/\mu_2^2$) given in the Table 1 are particularly useful for understanding the shape of the distribution.

For more detailed information on the Weibull distribution, see Kantar and Şenoğlu (2008).

In this study, we focus on the fiducial inference for testing the equality of μ_i's, i.e. the hypothesis in Eq. (4). For this purpose, we use the fiducial distribution proposed by Fisher (1930, 1933, 1935) to rectify the deficiency of the Bayesian approach, namely, a lack of enough prior knowledge on parameters (Cisewski and Hannig, 2012; Hannig, 2009). Fiducial distribution can be thought of as posterior distribution in the Bayesian framework without assuming a prior distribution on the parameters. Then, inferences are made about the parameters of interest by switching the role of the parameters and data (Hannig, 2009). Fiducial distribution for the parameter of interest is believed to have all of the information contained in the data about the parameter.

Fiducial inference has received considerable attention recently; therefore, we mention here some of the latest works related to it. Wang (2000) provided a general explanation of the fiducial confidence interval and explored a procedure for using the fiducial inference in evaluating the confidence coefficients of pre-data intervals. Hannig *et al.* (2006b) constructed fiducial generalized confidence intervals for ratio of means of two log-normal distributions. Hannig *et al.* (2006a) singled out a subclass of generalized pivotal quantities called fiducial generalized pivotal quantities (FGPQs). They showed that under some mild conditions, generalized confidence intervals constructed using FGPQs have correct frequentist coverage, at least asymptotically. O'Reilly and Rueda (2007) gave inference procedures both from the classical and the Bayesian view points. They showed numerically, through various examples, that the posterior distribution for the parameter and the induced fiducial distribution are almost equivalent. Lidong *et al.* (2008) developed a new technique based on fiducial argument for constructing confidence intervals for the variance components and the intraclass correlation in a two-component mixed-effects linear model. Hannig (2009) extended Fisher's fiducial argument and obtained a generalized fiducial recipe. Krishnamoorthy and Lee (2010) proposed a simple method of constructing confidence intervals for a function of binomial success probabilities and for a function of Poisson means. The method involves finding an approximate fiducial quantity for the parameters of interest. Li *et al.* (2011) proposed a fiducial-based test for the equality of several normal means when the variances are unknown and unequal. Wandler and Hannig (2011) proposed a solution to the inference on the largest mean of a multivariate normal distribution problem via fiducial inference methods. Wang *et al.* (2012) presented an approach based on the fiducial inference for constructing prediction intervals for any given distribution. Cisewski and Hannig (2012) presented an approach based on generalized fiducial inference for interval estimation for both balanced and unbalanced Gaussian linear mixed models. Zhao *et al.* (2012) introduced the fiducial empirical distribution of a random variable for nonparametric situations. Mathew and Young (2013) proposed fiducial quantities to construct approximate tolerance limits and intervals for functions of some discrete random variables. Li and Xu (2016) presented fiducial inference for the parameters of the Birnbaum–Saunders distribution. Yan and Xu (2017) constructed a fiducial generalized confidence interval which is proved to have correct asymptotic coverage for the slope parameter of the measurement error model with the reliability ratio known. Chen and Ye (2017) developed methods for constructing some important statistical limits of a gamma distribution. For each problem discussed in their study, the inferential procedure based on the generalized fiducial method was outlined and a simulation was conducted to assess its performance. Xu and Li (2018) presented a fiducial test for equality of regression coefficients in several normal models under heteroscedasticity. Eftekhar *et al.* (2018) considered the problem of testing the equality of several multivariate normal mean vectors under heteroscedasticity. Veronese and Melilli (2018) proposed a way to construct fiducial distributions for a multi-dimensional parameter.

To the best of our knowledge, this is the first study testing the equality of μ_i's when the scale parameters are heterogeneous in the context of fiducial inference.

The remainder of this chapter is organized as follows. Weibull distribution and the estimators of the μ_i and σ_i obtained by using MML methodology proposed by Tiku (1967) are presented in Section 2. The proposed test based on the fiducial inference is given in Section 3. The power performance of the tests under a variety of scenarios is discussed in Section 4. Concluding remarks are given in Section 5.

2.   Parameter Estimation

In this section, we obtain the estimators of μ_i and σ_i using the MML methodology.

Let $\{Y_{i1}, Y_{i2}, \ldots, Y_{in_i}\}$ be random samples from $Weibull(p, \mu_i, \sigma_i)$ $(i = 1, \ldots, a)$ distribution. The log-likelihood $(\ln L)$ function of the random sample is given as

$$\ln L = \sum_{i=1}^{a} n_i \ln\left(\frac{p}{\sigma_i}\right) + (p-1) \sum_{i=1}^{a} \sum_{j=1}^{n_i} \ln\left(\frac{y_{ij} - \mu_i}{\sigma_i}\right)$$
$$- \sum_{i=1}^{a} \sum_{j=1}^{n_i} \left(\frac{y_{ij} - \mu_i}{\sigma_i}\right)^p \quad (5)$$

The partial derivatives of $\ln L$ with respect to the parameters μ_i and σ_i are

$$\frac{\partial \ln L}{\partial \mu_i} = -\frac{(p-1)}{\sigma_i} \sum_{j=1}^{n_i} z_{ij}^{-1} + \frac{p}{\sigma_i} \sum_{j=1}^{n_i} z_{ij}^{p-1} = 0 \quad (6)$$

$$\frac{\partial \ln L}{\partial \sigma_i} = -\frac{n_i}{\sigma_i} - \frac{(p-1)}{\sigma_i} \sum_{j=1}^{n_i} z_{ij} z_{ij}^{-1} + \frac{p}{\sigma_i} \sum_{j=1}^{n_i} z_{ij} z_{ij}^{p-1} = 0 \quad (7)$$

where $z_{ij} = \frac{y_{ij} - \mu_i}{\sigma_i}$.

It is not possible to obtain closed-form expressions for the estimators of the parameters in Eqs. (6) and (7). We therefore resort to iterative methods to find their approximate solutions. However, iterative methods have some drawbacks, such as the risk of becoming trapped in false solutions and non-convergence of iterations. Therefore, in order to rule out problems caused by iterative methods, we utilize MML methodology providing explicit solutions for likelihood equations.

To obtain the MML estimators of the parameters, first we arrange $z_{ij}(i = 1, 2, \ldots, a; j = 1, 2, \ldots, n_i)$ in the ascending order, i.e. $z_{i(1)} \leq z_{i(2)} \leq \cdots \leq z_{i(n_i)}$ and rewrite the likelihood equations in terms of order statistics as follows:

$$\frac{\partial \ln L}{\partial \mu_i} = -\frac{(p-1)}{\sigma_i} \sum_{j=1}^{n_i} g_1(z_{i(j)}) + \frac{p}{\sigma_i} \sum_{j=1}^{n_i} g_2(z_{i(j)}) = 0 \qquad (8)$$

$$\frac{\partial \ln L}{\partial \sigma_i} = -\frac{n_i}{\sigma_i} - \frac{(p-1)}{\sigma_i} \sum_{j=1}^{n_i} z_{i(j)} g_1(z_{i(j)})$$

$$+ \frac{p}{\sigma_i} \sum_{j=1}^{n_i} z_{i(j)} g_2(z_{i(j)}) = 0 \qquad (9)$$

It is easy to verify that $\sum_{j=1}^{n_i} z_{i(j)} = \sum_{j=1}^{n_i} z_{ij}$. Here, $z_{i(j)} = \frac{y_{i(j)} - \mu_i}{\sigma_i}$, $g_1(z) = z^{-1}$ and $g_2(z) = z^{(p-1)}$.

Second, the first two terms of the Taylor series expansion about the expected values of the order statistics, i.e. $t_{i(j)} = E(z_{i(j)})$, is used to linearize the nonlinear functions $g_1(z_{i(j)})$ and $g_2(z_{i(j)})$ as

$$g_1(z_{i(j)}) = \alpha_{j0} - \beta_{j0} z_{i(j)} \qquad (10)$$

and

$$g_2(z_{i(j)}) = \alpha_{ij} + \beta_{ij} z_{i(j)} \qquad (11)$$

where $\alpha_{j0} = 2t_{i(j)}^{-1}$, $\beta_{j0} = t_{i(j)}^{-2}$, $\alpha_{ij} = (2-p)t_{i(j)}^{(p-1)}$ and $\beta_{ij} = (p-1)t_{i(j)}^{(p-2)}$.

It is hard to obtain exact values of $t_{i(j)}$; therefore, we use their approximate values obtained from the following equality:

$$\int_0^{t_{i(j)}} f(z) = \frac{j}{n_i + 1} \tag{12}$$

Note that these approximate values do not affect the performance of the proposed methodology adversely.

Finally, we plug linearized functions into Eqs. (8) and (9) and we get the following modified likelihood equations:

$$\frac{\partial \ln L^*}{\partial \mu_i} = -\frac{(p-1)}{\sigma_i} \sum_{j=1}^{n_i} (\alpha_{j0} - \beta_{j0} z_{i(j)})$$

$$+ \frac{p}{\sigma_i} \sum_{j=1}^{n_i} (\alpha_{ij} + \beta_{ij} z_{i(j)}) = 0 \tag{13}$$

$$\frac{\partial \ln L^*}{\partial \sigma_i} = -\frac{n_i}{\sigma_i} - \frac{(p-1)}{\sigma_i} \sum_{j=1}^{n_i} z_{i(j)} (\alpha_{j0} - \beta_{j0} z_{i(j)})$$

$$+ \frac{p}{\sigma_i} \sum_{j=1}^{n_i} z_{i(j)} (\alpha_{ij} + \beta_{ij} z_{i(j)}) = 0 \tag{14}$$

Following MML estimators are obtained by solving Eqs. (13) and (14) with respect to the parameters of interest:

$$\hat{\mu}_i = \hat{\mu}_n - \frac{\Delta_i}{m_i} \hat{\sigma}_i \quad \text{and} \quad \hat{\sigma}_i = \frac{-B_i + \sqrt{B_i^2 + 4A_i C_i}}{2\sqrt{A_i(A_i - 1)}} \tag{15}$$

where $A_i = n_i$, $B_i = \sum_{j=1}^{n_i} \Delta_j (y_{i(j)} - \hat{\mu}_n)$, $C_i = \sum_{j=1}^{n_i} \delta_j (y_{i(j)} - \hat{\mu}_n)^2$, $\delta_j = (p-1)\beta_{j0} + p\beta_{ij}$, $m_i = \sum_{j=1}^{n_i} \delta_j$, $\hat{\mu}_n = \sum_{j=1}^{n_i} \frac{\delta_j y_{i(j)}}{m_i}$, $\Delta_j = (p-1)\alpha_{j0} - p\alpha_{ij}$ and $\Delta_i = \sum_{j=1}^{n_i} \Delta_j$.

As aforementioned, the MML estimators have the following attractive properties:

(i) Contrary to ML estimators, MML estimators do not require any iterative estimation process because they are explicit functions of the sample observations.

(ii) Similar to ML estimators, MML estimators are asymptotically unbiased, efficient and consistent under regularity conditions.

(iii) MML estimators are as efficient as the ML estimators even for small sample sizes.

(iv) MML estimators are remarkably robust to departures from the assumed model.

It should be considered that if large samples ($n > 250$ or so) are not available, in case of simultaneous estimation, ML methods are of doubtful accuracy. However, in general, sample size is much smaller than 250 in the context of experimental design. Therefore, in this study, we concentrate on the two-parameter Weibull in place of the three-parameter Weibull, namely shape parameter p is assumed to be known. On the other hand, a reasonable value of the shape parameter can be estimated by using the profile likelihood methodology whose steps are given as follows:

Step 1: MML estimators of the model parameters, i.e. $\hat{\mu}_i$ and $\hat{\sigma}_i$ $(i = 1, 2, \ldots, a)$ are calculated for given value of the shape parameter p.

Step 2: $\ln L(\hat{\mu}_i, \hat{\sigma}_i, p)$ is computed.

Step 3: Steps (1) and (2) are repeated for a serious values of shape parameter p.

Step 4: Value of shape parameter maximizing the $\ln L$ is taken to be the estimate value for p.

MML methodology is known to be robust to the misspecification of the shape parameter p over a plausible range. Therefore, separate estimation of the shape parameter p does not affect the efficiencies of the MML estimators of μ_i and σ_i $(i = 1, 2, \ldots, a)$ negatively (Acıtaş *et al.*, 2013).

3. Proposed Test Based on Fiducial Inference

Theory of fiducial inference is given briefly as follows. Let $Y = G(\theta, U)$ be a data-generating equation or structural equation showing the relationship between the data and the parameters. Also, let $Q(y, u) = \{\theta : y = G(\theta, u)\}$ be a set-valued function which is defined as the inverse image of G. Here, Y is a random vector indexed by parameter(s) $\theta \in \Theta$, U is the random element whose distribution is independent of parameters and completely known, y is the observed value of Y and u is any realization of U. By means of given data and values which can be generated randomly for U, fiducial distribution of the parameter can be defined.

In the context of equality of μ_i's, when the distribution of Y_{ij}'s $(i = 1, 2, \ldots, a; j = 1, 2, \ldots, n_i)$ is normal and the scale parameters are known, the test statistics for Eq. (4) is

$$T(\bar{Y}; \sigma^2) = \sum_{i=1}^{a} \frac{n_i}{\sigma_i^2}(\bar{Y}_i - \mu_i)^2 - \frac{\left(\sum_{i=1}^{a} n_i(\bar{Y}_i - \mu_i)/\sigma_i^2\right)^2}{\sum_{i=1}^{a} n_i/\sigma_i^2} \tag{16}$$

where $\bar{Y}_i = \sum_{j=1}^{n_i} Y_{ij}/n_i$. Asymptotically, $T(\bar{Y}; \sigma^2) \sim \chi^2_{(a-1)}$.

If the scale parameters are unknown, the test statistics is

$$T(\bar{Y}; S^2) = \sum_{i=1}^{a} \frac{n_i}{S_i^2}(\bar{Y}_i - \mu_i)^2 - \frac{\left(\sum_{i=1}^{a} n_i(\bar{Y}_i - \mu_i)/S_i^2\right)^2}{\sum_{i=1}^{a} n_i/S_i^2} \tag{17}$$

Under H_0, the test statistics in Eq. (17) is written as follows:

$$T(\bar{Y}; S^2) = \sum_{i=1}^{a} \frac{n_i}{S_i^2}\bar{Y}_i^2 - \frac{\left(\sum_{i=1}^{a} n_i\bar{Y}_i/S_i^2\right)^2}{\sum_{i=1}^{a} n_i/S_i^2} \tag{18}$$

where $S_i^2 = \sum_{j=1}^{n_i}(Y_{ij} - \bar{Y}_i)^2/(n_i - 1)$ for $i = 1, \ldots, a$.

In the following, an attempt is made to propose a test statistics based on fiducial inference when the Y_{ij}'s follow Weibull distribution. In other words, the following test statistics is proposed as an alternative to the above test statistics when $\{Y_{i1}, Y_{i2}, \ldots, Y_{in_i}\} \sim Weibull(p, \mu_i, \sigma_i)$ $(i = 1, \ldots, a)$:

$$T_R^* = \sum_{i=1}^{a} \left(\frac{(\hat{\mu}_i - \mu_i) - \frac{\sum_{i=1}^{a}(\hat{\mu}_i - \mu_i)/V(\hat{\mu}_i)}{\sum_{i=1}^{a} 1/V(\hat{\mu}_i)}}{\sqrt{V(\hat{\mu}_i)}} \right)^2 \tag{19}$$

Under H_0, this can be reduced to

$$T_R^* = \sum_{i=1}^{a} \left(\frac{\hat{\mu}_i - \bar{\hat{\mu}}}{\sqrt{V(\hat{\mu}_i)}} \right)^2 \tag{20}$$

where $\bar{\hat{\mu}} = \frac{\sum_{i=1}^{a} w_i \hat{\mu}_i}{w}$, $w_i = \frac{1}{V(\hat{\mu}_i)}$ and $w = \sum_{i=1}^{a} w_i$.

In an effort to find the fiducial distribution of the test statistics in Eq. (19), we first use the following pivotal quantity:

$$Z_i = \frac{\hat{\mu}_i - \mu_i}{\sqrt{V(\hat{\mu}_i)}} \tag{21}$$

where Z_i is asymptotically distributed as $N(0,1)$ for $i = 1, \ldots, a$.

Structural equation obtained in terms of the pivotal quantity is

$$\hat{\mu}_i = \mu_i + Z_i \sqrt{V(\hat{\mu}_i)} \quad i = 1, \ldots, a \tag{22}$$

Given an observation $\hat{\mu}_i$, the solution of Eq. (22) can be written as follows:

$$\mu_i = \hat{\mu}_i - z_i \sqrt{V(\hat{\mu}_i)} \tag{23}$$

where z_i is a realization of Z_i.

Hence, the fiducial distribution of μ_i is the same as that of

$$T^*_{\mu_i} = \hat{\mu}_i - z_i \sqrt{V(\hat{\mu}_i)} \tag{24}$$

Using the fiducial distribution of μ_i in Eq. (24), the fiducial distribution of T^*_R in Eq. (19) is obtained as follows:

$$FT^*_R = \sum_{i=1}^{a} z_i^2 - \frac{\left(\sum_{i=1}^{a} z_i / \sqrt{V(\hat{\mu}_i)}\right)^2}{\sum_{i=1}^{a} 1/V(\hat{\mu}_i)} \tag{25}$$

The corresponding p-value for the null hypothesis in Eq. (4) is defined by

$$p = P(FT^*_R > T^*_{R0} | H_0) \tag{26}$$

where T^*_{R0} is the observed value of T^*_R in Eq. (20). H_0 is rejected if the p-value is less than the presumed significance level α, i.e. $p < \alpha$.

The following algorithm is used to compute the simulated value of the fiducial p-value for the proposed test:

Step 1: Compute $T^*_R = \sum_{i=1}^{a} \left(\frac{\hat{\mu}_i - \hat{\bar{\mu}}}{\sqrt{V(\hat{\mu}_i)}} \right)^2$ for a given value of $\hat{\mu}_i$ $(i = 1, \ldots, a)$.

Step 2: For $j = 1$ to M,

— generate $z_i \sim N(0, 1)$, $\quad i = 1, \ldots, a$.

— compute $FT_R^* = \sum_{i=1}^{a} z_i^2 - \dfrac{\left(\sum_{i=1}^{a} z_i / \sqrt{V(\hat{\mu}_i)}\right)^2}{\sum_{i=1}^{a} 1/V(\hat{\mu}_i)}$.

— if $FT_R^* > T_{R0}^*$, set $W_j = 1$, else $W_j = 0$ (end loop).

Step 3: Compute p-value using the equality $p = \frac{1}{M} \sum_{i=1}^{M} W_i$.

See also Güven *et al.* (2019) and the references given there.

4. Simulation Study

In this section, Monte Carlo simulation study is carried out to assess the performances of the proposed robust fiducial p-value test (RFT) based on MML estimators and fiducial p-value test (FT) based on corrected least squares (LS) estimators. FT test is defined by replacing $\hat{\mu}_i$ in FT_R^* with the corrected LS estimator $\tilde{\mu}_i$. Here,

$$\tilde{\mu}_i = \bar{Y}_i - \Gamma\left(1 + \frac{1}{p}\right)\tilde{\sigma}_i \qquad (27)$$

where $\tilde{\sigma}_i = \sqrt{S_i^2 \Big/ \left(\Gamma\left(1 + \frac{2}{p}\right) - \Gamma^2\left(1 + \frac{1}{p}\right)\right)}$, $S_i^2 = \sum_{j=1}^{n_i}(Y_{ij} - \bar{Y}_i)^2/$
$(n_i - 1)$ and $\bar{Y}_i = \sum_{j=1}^{n_i} Y_{ij}/n_i$ for $i = 1, \ldots, a$; $j = 1, 2, \ldots, n_i$.

Simulated type I errors and power values of RFT and FT tests are compared at significance level $\alpha = 0.050$ for different configurations of sample sizes (n), shape parameters (p) and scale parameters (σ^2). They are given as follows:

$$n = (4, 4, 4), (4, 5, 6), (4, 6, 8)$$
$$p = 2.5, 3, 3.5, 4, 6$$
$$\sigma^2 = (1, 1, 1), (1, 1.5, 2), (1, 2, 4)$$

Data are generated from the Weibull distribution and the observed value of T_R^* in Eq. (20), namely T_{R0}^*, is calculated for a given value of $\hat{\mu}_i$. Then, a random number from standard normal distribution is generated for each group and FT_R^* value is computed accordingly for 10,000 iterations. The p-value for the RFT test is obtained by calculating the proportion of cases in which FT_R^* value exceeds T_{R0}^*. This simulation is run 10,000 times in order to obtain the type I errors by counting the number of p-values less than the significance level of $\alpha = 0.050$. The simulated type I errors of the *RFT* and *FT* tests are shown in Table 2.

It is seen from Table 2 that the simulated type I error values do not depart from the presumed significance level for all cases. In other words, irrespective of the settings of parameter values and sample sizes, the simulated type I error values of *RFT* and *FT* tests are very close to $\alpha = 0.050$. Any test can be considered powerful if it achieves maximum power and adheres to the prescribed significance level. The simulated powers of the tests are investigated under the scenarios mentioned earlier and they are presented in Tables 3–7.

Difference among the treatment means is formed by adding a constant d to the observations in the first treatment and subtracting a constant d from the observations in the third treatment. Consequently, the first line of Tables 3–7, i.e. $d = 0.00$, presents simulated type I errors of *RFT* and *FT* tests.

Tables 3–7 show that the *RFT* test outperforms the *FT* test for all settings of shape parameter p. It is obvious that the *RFT* test shows superior performance than the *FT* test, especially for $p = 2.5$ and 3. As p increases, the powers of the two tests get closer to each other, but the *RFT* test is still slightly better than the *FT* test. It should also be noted that as the between group variance disparity increases, the *RFT* test performs better than the *FT* test.

Robustness: In practice, it is not always possible to correctly specify a model. Therefore, it is essential to investigate the performance of a statistical test under the presence of some misspecifications. A test is regarded as robust if it achieves nominal type I error rates and high power values simultaneously under departures from an underlying distribution. In this study, we assume that the true model is $Weibull(p = 3.5, \sigma)$ and consider the following Huber's contamination models as plausible alternatives:

Model 1. 0.90 Weibull $(p = 3.5, \sigma) + 0.10$ Weibull $(p = 2, \sigma)$
Model 2. 0.80 Weibull $(p = 3.5, \sigma) + 0.20$ Weibull $(p = 2, \sigma)$

Here, 0.10 and 0.20 are the expected fractions of adversarial outliers, namely 0.10 and 0.20 fractions of observations are subject to arbitrary adversarial noise.

Table 2. Simulated type I errors of RFT and FT tests.

	$p = 2.5$		$p = 3$		$p = 3.5$		$p = 4$		$p = 6$	
	RFT	FT	RFT	FT	RFT	FT	RFT	FT	RFT	FT
$n = (4,4,4)$										
$\sigma^2 = (1,1,1)$	0.050	0.051	0.046	0.047	0.048	0.048	0.046	0.045	0.051	0.051
$\sigma^2 = (1,1.5,2)$	0.049	0.049	0.047	0.046	0.044	0.044	0.050	0.051	0.050	0.050
$\sigma^2 = (1,2,4)$	0.051	0.048	0.047	0.046	0.048	0.046	0.046	0.045	0.047	0.048
$n = (4,5,6)$										
$\sigma^2 = (1,1,1)$	0.050	0.051	0.050	0.051	0.048	0.047	0.048	0.047	0.048	0.048
$\sigma^2 = (1,1.5,2)$	0.051	0.052	0.048	0.048	0.047	0.046	0.045	0.045	0.050	0.051
$\sigma^2 = (1,2,4)$	0.051	0.051	0.048	0.046	0.050	0.049	0.047	0.047	0.048	0.048
$n = (4,6,8)$										
$\sigma^2 = (1,1,1)$	0.051	0.050	0.047	0.047	0.048	0.048	0.048	0.047	0.045	0.047
$\sigma^2 = (1,1.5,2)$	0.048	0.049	0.048	0.047	0.047	0.047	0.047	0.046	0.051	0.050
$\sigma^2 = (1,2,4)$	0.048	0.049	0.050	0.050	0.047	0.047	0.047	0.046	0.050	0.050

Table 3.   Simulated powers of RFT and FTs tests for $p = 2.5$.

$n = (4,4,4)$			$n = (4,4,4)$			$nn = (4,4,4)$		
$\sigma^2 = (1,1,1)$			$\sigma^2 = (1,1.5,2)$			$\sigma^2 = (1,2,4)$		
d	RFT	FT	d	RFT	FT	d	RFT	FT
0.00	0.050	0.051	0.00	0.049	0.049	0.00	0.051	0.048
0.20	0.12	0.10	0.24	0.15	0.11	0.28	0.16	0.11
0.40	0.40	0.30	0.48	0.39	0.30	0.56	0.40	0.29
0.60	0.68	0.58	0.72	0.72	0.60	0.84	0.72	0.58
0.80	0.91	0.84	0.96	0.93	0.85	1.12	0.92	0.82
1.00	0.99	0.96	1.20	0.99	0.96	1.40	0.99	0.96
$n = (4,5,6)$			$n = (4,5,6)$			$n = (4,5,6)$		
$\sigma^2 = (1,1,1)$			$\sigma^2 = (1,1.5,2)$			$\sigma^2 = (1,2,4)$		
d	RFT	FT	d	RFT	FT	d	RFT	FT
0.00	0.050	0.051	0.00	0.051	0.052	0.00	0.051	0.051
0.20	0.12	0.11	0.20	0.11	0.10	0.24	0.13	0.10
0.40	0.39	0.33	0.40	0.32	0.25	0.48	0.36	0.26
0.60	0.76	0.67	0.60	0.63	0.52	0.72	0.66	0.53
0.80	0.95	0.90	0.80	0.88	0.79	0.96	0.90	0.79
1.00	0.99	0.98	1.00	0.98	0.94	1.20	0.98	0.94
$n = (4,6,8)$			$n = (4,6,8)$			$n = (4,6,8)$		
$\sigma^2 = (1,1,1)$			$\sigma^2 = (1,1.5,2)$			$\sigma^2 = (1,2,4)$		
d	RFT	FT	d	RFT	FT	d	RFT	FT
0.00	0.051	0.050	0.00	0.048	0.049	0.00	0.048	0.049
0.20	0.12	0.12	0.20	0.11	0.10	0.24	0.14	0.10
0.40	0.43	0.37	0.40	0.35	0.28	0.48	0.39	0.30
0.60	0.80	0.72	0.60	0.68	0.58	0.72	0.73	0.60
0.80	0.97	0.94	0.80	0.92	0.85	0.96	0.94	0.86
1.00	0.99	0.99	1.00	0.99	0.97	1.20	0.99	0.97

Table 4.   Simulated powers of *RFT* and *FT* tests for $p = 3$.

$n = (4,4,4)$			$n = (4,4,4)$			$n = (4,4,4)$		
$\sigma^2 = (1,1,1)$			$\sigma^2 = (1,1.5,2)$			$\sigma^2 = (1,2,4)$		
d	*RFT*	*FT*	d	*RFT*	*FT*	d	*RFT*	*FT*
0.00	0.046	0.047	0.00	0.047	0.046	0.00	0.047	0.046
0.20	0.11	0.10	0.24	0.12	0.11	0.28	0.13	0.10
0.40	0.32	0.28	0.48	0.34	0.28	0.56	0.35	0.28
0.60	0.63	0.57	0.72	0.65	0.57	0.84	0.64	0.55
0.80	0.86	0.81	0.96	0.89	0.83	1.12	0.88	0.80
1.00	0.97	0.95	1.20	0.98	0.96	1.40	0.97	0.94
$n = (4,5,6)$			$n = (4,5,6)$			$n = (4,5,6)$		
$\sigma^2 = (1,1,1)$			$\sigma^2 = (1,1.5,2)$			$\sigma^2 = (1,2,4)$		
d	*RFT*	*FT*	d	*RFT*	*FT*	d	*RFT*	*FT*
0.00	0.050	0.051	0.00	0.048	0.048	0.00	0.048	0.046
0.20	0.11	0.11	0.20	0.11	0.10	0.24	0.12	0.10
0.40	0.36	0.32	0.40	0.28	0.24	0.48	0.32	0.26
0.60	0.70	0.64	0.60	0.57	0.50	0.72	0.60	0.51
0.80	0.92	0.89	0.80	0.83	0.77	0.96	0.85	0.77
1.00	0.99	0.98	1.00	0.96	0.93	1.20	0.96	0.93
$n = (4,6,8)$			$n = (4,6,8)$			$n = (4,6,8)$		
$\sigma^2 = (1,1,1)$			$\sigma^2 = (1,1.5,2)$			$\sigma^2 = (1,2,4)$		
d	*RFT*	*FT*	d	*RFT*	*FT*	d	*RFT*	*FT*
0.00	0.047	0.047	0.00	0.048	0.047	0.00	0.050	0.050
0.20	0.11	0.11	0.20	0.10	0.09	0.24	0.12	0.10
0.40	0.38	0.34	0.40	0.32	0.28	0.48	0.34	0.28
0.60	0.75	0.70	0.60	0.63	0.57	0.72	0.66	0.58
0.80	0.96	0.93	0.80	0.87	0.82	0.96	0.90	0.84
1.00	0.99	0.99	1.00	0.98	0.96	1.20	0.98	0.96

Table 5. Simulated powers of RFT and FT tests for $p = 3.5$.

$n = (4, 4, 4)$			$n = (4, 4, 4)$			$n = (4, 4, 4)$		
$\sigma^2 = (1, 1, 1)$			$\sigma^2 = (1, 1.5, 2)$			$\sigma^2 = (1, 2, 4)$		
d	RFT	FT	d	RFT	FT	d	RFT	FT
0.00	0.048	0.048	0.00	0.044	0.044	0.00	0.048	0.046
0.20	0.10	0.10	0.24	0.11	0.10	0.32	0.14	0.12
0.40	0.29	0.27	0.48	0.31	0.27	0.64	0.40	0.34
0.60	0.58	0.54	0.72	0.62	0.56	0.96	0.71	0.64
0.80	0.84	0.80	0.96	0.85	0.80	1.28	0.92	0.88
1.00	0.96	0.94	1.20	0.97	0.95	1.60	0.99	0.98
$n = (4, 5, 6)$			$n = (4, 5, 6)$			$n = (4, 5, 6)$		
$\sigma^2 = (1, 1, 1)$			$\sigma^2 = (1, 1.5, 2)$			$\sigma^2 = (1, 2, 4)$		
d	RFT	FT	d	RFT	FT	d	RFT	FT
0.00	0.048	0.047	0.00	0.047	0.046	0.00	0.050	0.049
0.20	0.11	0.11	0.24	0.12	0.11	0.28	0.12	0.11
0.40	0.33	0.31	0.48	0.35	0.31	0.56	0.36	0.32
0.60	0.66	0.62	0.72	0.69	0.64	0.84	0.68	0.62
0.80	0.90	0.87	0.96	0.92	0.88	1.12	0.91	0.87
1.00	0.99	0.98	1.20	0.99	0.98	1.40	0.99	0.98
$n = (4, 6, 8)$			$n = (4, 6, 8)$			$n = (4, 6, 8)$		
$\sigma^2 = (1, 1, 1)$			$\sigma^2 = (1, 1.5, 2)$			$\sigma^2 = (1, 2, 4)$		
d	RFT	FT	d	RFT	FT	d	RFT	FT
0.00	0.048	0.048	0.00	0.047	0.047	0.00	0.047	0.047
0.20	0.11	0.11	0.24	0.13	0.12	0.24	0.11	0.10
0.40	0.36	0.34	0.48	0.40	0.36	0.48	0.32	0.28
0.60	0.71	0.67	0.72	0.75	0.71	0.72	0.62	0.56
0.80	0.93	0.91	0.96	0.95	0.94	0.96	0.86	0.82
1.00	0.99	0.99	1.20	0.99	0.99	1.20	0.97	0.96

Table 6. Simulated powers of RFT and FT tests for $p = 4$.

$n = (4,4,4)$			$n = (4,4,4)$			$n = (4,4,4)$		
$\sigma^2 = (1,1,1)$			$\sigma^2 = (1,1.5,2)$			$\sigma^2 = (1,2,4)$		
d	RFT	FT	d	RFT	FT	d	RFT	FT
0.00	0.046	0.045	0.00	0.050	0.051	0.00	0.046	0.045
0.20	0.10	0.10	0.24	0.11	0.10	0.28	0.11	0.10
0.40	0.28	0.26	0.48	0.30	0.27	0.56	0.29	0.25
0.60	0.55	0.52	0.72	0.58	0.54	0.84	0.57	0.53
0.80	0.80	0.77	0.96	0.82	0.79	1.12	0.81	0.77
1.00	0.95	0.93	1.20	0.96	0.94	1.40	0.95	0.93
$n = (4,5,6)$			$n = (4,5,6)$			$n = (4,5,6)$		
$\sigma^2 = (1,1,1)$			$\sigma^2 = (1,1.5,2)$			$\sigma^2 = (1,2,4)$		
d	RFT	FT	d	RFT	FT	d	RFT	FT
0.00	0.048	0.047	0.00	0.045	0.045	0.00	0.047	0.047
0.20	0.10	0.10	0.24	0.12	0.11	0.28	0.12	0.11
0.40	0.31	0.29	0.48	0.34	0.32	0.56	0.34	0.31
0.60	0.63	0.60	0.72	0.66	0.63	0.84	0.65	0.60
0.80	0.88	0.85	0.96	0.90	0.88	1.12	0.88	0.85
1.00	0.98	0.97	1.20	0.99	0.98	1.40	0.98	0.97
$n = (4,6,8)$			$n = (4,6,8)$			$n = (4,6,8)$		
$\sigma^2 = (1,1,1)$			$\sigma^2 = (1,1.5,2)$			$\sigma^2 = (1,2,4)$		
d	RFT	FT	d	RFT	FT	d	RFT	FT
0.00	0.048	0.047	0.00	0.047	0.046	0.00	0.047	0.046
0.20	0.11	0.10	0.24	0.12	0.11	0.24	0.11	0.10
0.40	0.35	0.33	0.48	0.37	0.34	0.48	0.30	0.27
0.60	0.68	0.65	0.72	0.72	0.69	0.72	0.58	0.54
0.80	0.92	0.90	0.96	0.94	0.92	0.96	0.83	0.80
1.00	0.99	0.98	1.20	0.99	0.99	1.20	0.96	0.95

Table 7.　Simulated powers of RFT and FT tests for $p = 6$.

$n = (4,4,4)$			$n = (4,4,4)$			$n = (4,4,4)$		
$\sigma^2 = (1,1,1)$			$\sigma^2 = (1,1.5,2)$			$\sigma^2 = (1,2,4)$		
d	RFT	FT	d	RFT	FT	d	RFT	FT
0.00	0.051	0.051	0.00	0.050	0.050	0.00	0.047	0.048
0.24	0.11	0.11	0.28	0.11	0.11	0.32	0.11	0.10
0.48	0.32	0.31	0.56	0.33	0.31	0.64	0.31	0.30
0.72	0.64	0.63	0.84	0.65	0.64	0.96	0.61	0.59
0.96	0.90	0.88	1.12	0.88	0.87	1.28	0.85	0.84
1.20	0.97	0.97	1.40	0.98	0.97	1.60	0.96	0.96
$n = (4,5,6)$			$n = (4,5,6)$			$n = (4,5,6)$		
$\sigma^2 = (1,1,1)$			$\sigma^2 = (1,1.5,2)$			$\sigma^2 = (1,2,4)$		
d	RFT	FT	d	RFT	FT	d	RFT	FT
0.00	0.048	0.048	0.00	0.050	0.051	0.00	0.048	0.048
0.20	0.10	0.10	0.24	0.10	0.10	0.28	0.10	0.10
0.40	0.27	0.26	0.48	0.29	0.28	0.56	0.29	0.27
0.60	0.55	0.53	0.72	0.59	0.57	0.84	0.58	0.56
0.80	0.82	0.81	0.96	0.85	0.84	1.12	0.83	0.82
1.00	0.96	0.95	1.20	0.97	0.96	1.40	0.96	0.95
$n = (4,6,8)$			$n = (4,6,8)$			$n = (4,6,8)$		
$\sigma^2 = (1,1,1)$			$\sigma^2 = (1,1.5,2)$			$\sigma^2 = (1,2,4)$		
d	RFT	FT	d	RFT	FT	d	RFT	FT
0.00	0.045	0.047	0.00	0.051	0.050	0.00	0.050	0.050
0.20	0.10	0.10	0.24	0.11	0.10	0.28	0.11	0.11
0.40	0.30	0.29	0.48	0.33	0.31	0.56	0.32	0.30
0.60	0.61	0.60	0.72	0.64	0.62	0.84	0.64	0.62
0.80	0.87	0.86	0.96	0.90	0.89	1.12	0.88	0.87
1.00	0.98	0.98	1.20	0.99	0.98	1.40	0.98	0.98

It is seen from Table 8 that the first line represents the simulated type I errors and the rest indicate power values as mentioned before. The RFT test performs much better than the FT test for both Model I and Model II, especially when the heterogeneity in

Table 8. Simulated powers of RFT and FT tests for the models.

				Model I				
$n=(4,4,4)$			$n=(4,4,4)$			$n=(4,5,6)$		
$\sigma^2=(1,1.5,2)$			$\sigma^2=(1,2,4)$			$\sigma^2=(1,1.5,2)$		
d	RFT	FT	d	RFT	FT	d	RFT	FT
0.00	0.049	0.049	0.00	0.047	0.047	0.00	0.048	0.048
0.28	0.11	0.10	0.32	0.11	0.09	0.24	0.10	0.09
0.56	0.34	0.30	0.64	0.31	0.26	0.48	0.27	0.25
0.84	0.66	0.61	0.96	0.61	0.54	0.72	0.58	0.53
1.12	0.89	0.84	1.28	0.85	0.80	0.96	0.85	0.81
1.40	0.98	0.97	1.60	0.97	0.94	1.20	0.97	0.95
$n=(4,5,6)$			$n=(4,6,8)$			$n=(4,6,8)$		
$\sigma^2=(1,2,4)$			$\sigma^2=(1,1.5,2)$			$\sigma^2=(1,2,4)$		
d	RFT	FT	d	RFT	FT	d	RFT	FT
0.00	0.051	0.050	0.00	0.050	0.050	0.00	0.051	0.051
0.28	0.10	0.09	0.24	0.10	0.10	0.28	0.10	0.09
0.56	0.28	0.25	0.48	0.31	0.29	0.56	0.30	0.27
0.84	0.58	0.52	0.72	0.63	0.59	0.84	0.63	0.58
1.12	0.83	0.78	0.96	0.89	0.86	1.12	0.88	0.84
1.40	0.96	0.93	1.20	0.98	0.97	1.40	0.98	0.96
				Model II				
$n=(4,4,4)$			$n=(4,4,4)$			$n=(4,5,6)$		
$\sigma^2=(1,1.5,2)$			$\sigma^2=(1,2,4)$			$\sigma^2=(1,1.5,2)$		
d	RFT	FT	d	RFT	FT	d	RFT	FT
0.00	0.052	0.052	0.00	0.048	0.049	0.00	0.049	0.050
0.28	0.10	0.09	0.32	0.09	0.08	0.28	0.10	0.09
0.56	0.29	0.25	0.64	0.25	0.21	0.56	0.31	0.28
0.84	0.59	0.53	0.96	0.52	0.46	0.84	0.64	0.60
1.12	0.83	0.78	1.28	0.79	0.73	1.12	0.90	0.87
1.40	0.96	0.94	1.60	0.94	0.91	1.40	0.98	0.97

(*Continued*)

Table 8. (*Continued*)

$n = (4, 5, 6)$			$n = (4, 6, 8)$			$n = (4, 6, 8)$		
$\sigma^2 = (1, 2, 4)$			$\sigma^2 = (1, 1.5, 2)$			$\sigma^2 = (1, 2, 4)$		
d	*RFT*	*FT*	d	*RFT*	*FT*	d	*RFT*	*FT*
0.00	0.052	0.053	0.00	0.051	0.051	0.00	0.052	0.052
0.32	0.10	0.09	0.24	0.09	0.08	0.28	0.08	0.08
0.64	0.27	0.24	0.48	0.25	0.23	0.56	0.23	0.20
0.96	0.61	0.55	0.72	0.56	0.52	0.84	0.52	0.47
1.28	0.87	0.82	0.96	0.82	0.79	1.12	0.79	0.75
1.60	0.97	0.96	1.20	0.96	0.85	1.40	0.95	0.93

variances increases. Therefore, for contamination models, the *RFT* test achieves statistical robustness.

5. Conclusion

Reviewing the literature showed that comparing the equality of μ_i's is a scientific question of interest, especially when the usual normality and homogeneity of variances assumptions are violated. For this reason, in this study, we proposed a new test based on Fisher's fiducial inference for the equality of location parameters of Weibull populations under the heterogeneity of scale parameters assumption. In the process of estimation, the MML methodology which is asymptotically equivalent to the ML methodology is used. The simulation results show that the proposed test is preferred to the corresponding test based on the corrected LS estimators in terms of power criterion in all scenarios. It is also demonstrated that the proposed test is highly robust against contamination. As far as we know, there are no previous works on fiducial inference in the context of testing the equality of the location parameters of Weibull populations when the scale parameters are heterogeneous.

References

Acıtaş, Ş., Kasap, P., Şenoğlu, B. and Arslan, O. (2013). One-step *M*-estimators: Jones and Faddy's skewed *t*-distribution. *Journal of Applied Statistics*, **40**(7): 1545–1560.

Acıtaş, Ş., Aladağ, Ç.H. and Şenoğlu, B. (2019). A new approach for estimating the parameters of Weibull distribution via particle swarm optimization: An application to the strengths of glass fibre data. *Reliability Engineering and System Safety*, **183**: 116–127.

Chen, P. and Ye, Z.S. (2017). Approximate statistical limits for a gamma distribution. *Journal of Quality Technology*, **49**(1): 64–77.

Cisewski, J. and Hannig, J. (2012). Generalized fiducial inference for normal linear mixed models. *Annals of Statistics*, **40**(4): 2102–2127.

Cohen, C.A. and Whitten, B. (1982). Modified maximum likelihood and modified moment estimators for the three parameter Weibull distribution. *Communications in Statistics — Theory and Methods*, **11**(23): 2631–2656.

Eftekhar, S., Sadooghi-Alvandi, M. and Kharrati-Kopaei, M. (2018). Testing the equality of several multivariate normal mean vectors under heteroscedasticity: A fiducial approach and an approximate test. *Communications in Statistics — Theory and Methods*, **47**(7): 1747–1766.

Fisher, R.A. (1930). Inverse probability. *Mathematical Proceedings of the Cambridge. Cambridge University Press*, **26**(4): 528–535.

Fisher, R.A. (1933). The concepts of inverse probability and fiducial probability referring to unknown parameters. *Proceedings of the Royal Society of London A*, **139**(838): 343–348.

Fisher, R.A. (1935). The fiducial argument in statistical inference. *Ann Eugen*, **6**(4): 391–398.

Güven, G., Gürer, Ö., Şamkar, H. and Şenoğlu, B. (2019). A fiducial-based approach to the one-way ANOVA in the presence of nonnormality and heterogeneous error variances. *Journal of Statistical Computation and Simulation*, **89**(9): 1715–1729.

Hannig, J. (2009). On generalized fiducial inference. *Statistica Sinica*, **19**(2): 491–544.

Hannig, J., Iyer, H. and Patterson, P. (2006a). Fiducial generalized confidence intervals. *Journal of the American Statistical Association*, **101**(473): 254–269.

Hannig, J., Lidong, E., Abdel-Karim, A. and Iyer, H. (2006b). Simultaneous fiducial generalized confidence intervals for ratios of means of lognormal distributions. *Austrian Journal of Statistics*, **35**(2,3): 261–269.

Kantar, Y.M. and Şenoğlu, B. (2008). A comparative study for the location and scale parameters of the Weibull distribution with given shape parameter. *Computers and & Geosciences*, **34**(12): 1900–1909.

Krishnamoorthy, K. and Lee, M. (2010). Inference for functions of parameters in discrete distributions based on fiducial approach: Binomial and Poisson cases. *Journal of Statistical Planning and Inference*, **140**(5): 1182–1192.

Lai, C.D., Murthy, D.N. and Xie, M. (2006). Weibull distributions and their applications. In *Springer Handbook of Engineering Statistics*, pp. 63–78 (Springer, London).

Li, X., Wang, J. and Liang, H. (2011). Comparison of several means: A fiducial based approach. *Computational Statistics and Data Analysis*, **55**(5): 1993–2002.

Li, Y. and Xu, A. (2016). Fiducial inference for Birnbaum-Saunders distribution. *Journal of Statistical Computation and Simulation*, **86**(9): 1673–1685.

Lidong, E., Hannig, J. and Iyer, H. (2008). Fiducial intervals for variance components in an unbalanced two-component normal mixed linear model. *Journal of the American Statistical Association*, **103**(482): 854–865.

Mathew, T. and Young, D.S. (2013). Fiducial-based tolerance intervals for some discrete distributions. *Computational Statistics and Data Analysis*, **61**: 38–49.

O'Reilly, F. and Rueda, R. (2007). Fiducial inferences for the truncated exponential distribution. *Communications in Statistics — Theory and Methods*, **36**(12): 2207–2212.

Tiku, M.L. (1967). Estimating the mean and standard deviation from a censored normal sample. *Biometrika*, **54**(1–2): 155–165.

Veronese, P. and Melilli, E. (2018). Fiducial, confidence and objective Bayesian posterior distributions for a multidimensional parameter. *Journal of Statistical Planning and Inference*, **195**: 153–173.

Wandler, D.V. and Hannig, J. (2011). Fiducial inference on the largest mean of a multivariate normal distribution. *Journal of Multivariate Analysis*, **102**(1): 87–104.

Wang, Y.H. (2000). Fiducial intervals: What are they? *American Statistical*, **54**(2): 105–111.

Wang, C.M., Hannig, J. and Iyer, H.K. (2012). Fiducial prediction intervals. *Journal of Statistical Planning and Inference*, **142**(7): 1980–1990.

Xu, J. and Li, X. (2018). A fiducial *p*-value approach for comparing heteroscedastic regression models. *Communications in Statistics — Simulation and Computation*, **47**(2): 420–431.

Yan, L. and Xu, X. (2017). A new confidence interval in measurement error model with the reliability ratio known. *Communications in Statistics — Theory and Methods*, **46**(19): 9636–9650.

Zhao, S., Xu, X. and Ding, X. (2012). Fiducial inference under nonparametric situations. *Journal of Statistical Planning and Inference*, **142**(10): 2779–2798.

Chapter 6

Understanding the Effects of Green Swan Events on Financial Stability: Annex II Countries and Turkey

Şenay Açıkgöz[†,§] and Şahika Gökmen[*,†,‡,¶]

[†]*Department of Econometrics,
Ankara Hacı Bayram Veli University, Turkey*
[‡]*Department of Statistics, Uppsala University, Sweden*
[§]*senay.acikgoz@hbv.edu.tr*
[¶]*sahika.gokmen@statistics.uu.se*

Green swan events are as important and unpredictable as black swan events since they include climate-related risks and their consequences. One of the important consequences of such events is related to financial risks, such as credit, market, operational, liquidity and insurance risks, that create financial and price instability. Therefore, almost all central banks started to have a role in fighting climate change to maintain financial stability — the main aim of these banks. The relationship between the building blocks of green swan events and financial stability is constructed as the main idea of this chapter. Based on this, this research summarizes the mechanisms between climate-related financial risks and financial stability as well as price stability in the green swan context. The current situation of the world and Turkey are reviewed with data, and the interaction between climate risks and financial and price stabilities is explained. A couple of stability indicators for Annex II countries and

---

[*]Corresponding author.

Turkey are also reviewed to understand their current financial stability situation. Scenario analysis, which is the main tool for understanding the possible effects of climate change to produce policies for transition to low-carbon or net-zero target, is also reviewed. Therefore, this chapter contributes to this very recent literature by giving a clear view of the green swan.

1.   Introduction

As a legally binding international treaty on climate change, The Paris Agreement aims to limit global warming to well below 2°C (preferably 1.5°C) compared to the pre-industrial level. It is a long-term temperature goal, and countries have taken responsibility for reaching this goal in the fight against climate change. This responsibility brings in its wake that which requires a major shift in investment patterns toward low-carbon, climate-resilient options. Financing has become more critical than ever in order to achieve this goal by changing the regime with policies involving economic, social, technological, transformation and even for health systems because of possible future pandemics. An economy is stable if it is resilient against unexpected circumstances that lead to a flaw in the balance of the financial system. The relevant literature has observed that countries will have financial instabilities now and in the future due to climate-related risks.

Green swan events, which are derived from the definition of black swan, include the climate-related risk and its consequences. According to Taleb (2007), black swans are important for understanding the complex economic systems and green swans include all the climate-related risks, starting from natural disasters (including the pandemic) to big financial problems. In this way, climate change can affect financial stability through physical and transition risks that are defined under green swan. The first is the risks arising from the interaction of climate-related hazards with the vulnerability of exposure to human and natural systems; the latter is the risks associated with the uncertain financial impacts that could result from a rapid low-carbon transition, including policy changes, reputational impacts, technological breakthroughs or limitations and shifts in market preferences and social norms (Bolton *et al.*, 2020a). These green swan events can be related to financial risks such as credit, market, operational, liquidity and insurance risks (DG *et al.*, 2017). As noted in Carney (2015), one of the tasks of central banks and financial institutions is, therefore, to ensure that climate-related risks become integrated into financial stability monitoring and prudential supervision.

Financial stability, which significantly increases the effectiveness of the monetary policy, creates adverse effects on a country's economy and social welfare if it cannot be achieved. Therefore, all central banks around the world follow the stability of the financial system very closely and it is at the top of the agenda of central banks together with price stability. Price stability matters in the fight against climate change because it has a direct (arising from various environmental shifts, such as hotter temperatures, rising sea levels and more frequent and extreme storms, floods and droughts) and indirect (arising from attempts to adapt to these new conditions and from efforts to limit or mitigate climate change through a transition to a low-carbon economy) effects on the economies (Rudebusch, 2019). Therefore, climate change will entail broad structural changes and poses considerable financial risks for private households, businesses and financial institutions (Bremus *et al.*, 2020).

The world has already started taking action to fight against climate change and created relevant institutions or organizations. For example, the Financial Stability Board (FSB) — an international body that monitors and makes recommendations about the global financial system — established the Task Force on Climate-related Financial Disclosures (TCDF) to develop voluntary, consistent climate-related financial risk disclosures for use by companies in providing information to lenders, insurers, investors and other stakeholders. It is aimed to provide more transparent information flow among the economic agents to understand the risks posed by climate change. The final goal is to make more efficient markets and more stable and resilient economies in financially achieving this combat. This Task Force also recommends that organizations describe the resilience of their strategy, taking into consideration different climate-related scenarios, including a 2°C or lower scenario, where such information is material. Another organization is the Central Banks and Supervisors Network for Greening the Financial System (NGFS) launched by the Banque de France at the end of 2017. The aim of NGFS, which includes 65 central banks, is to enhance the role of the financial system to manage risks and to mobilize capital for green and low-carbon investments.[1]

Climate-related risks can be evaluated both at the micro and macro levels. At the micro level, financial institutions provide services for the firms in their transition with funding. At the macro level, not only the decrease in these risks but also some other achievements will be desired. Sustainable growth is one of them and the transition to a low-carbon economy might cause some harm before reaching its target because net-zero emissions require switching to renewable energy. Therefore, scenario analysis also needs to be done at a macro level to give insights to governments and their policymakers. Financial and price stability conditions can be assessed in such an analysis.

The related literature has mostly focused on the financial effects of black swan events (Brunaker and Nordqvist, 2013; Higgins 2014; Gocen *et al.*, 2015; Lleo and Ziemba, 2015; Phan and Wood, 2020; Simianer and Reimer, 2021). Such analyses for green swan events and their relationship with financial stability are rare since they recently entered the literature (Bolton *et al.*, 2020a). However, the climate risks and financial stability have been discussed more often (e.g. Say and Yucel, 2006; Halicioglu 2009; Bozkurt and Akan, 2014; Gökmenoglu and Taspinar, 2016; Steininger *et al.*, 2016; Nasreen *et al.*, 2017; Campiglio *et al.*, 2018; Brainard, 2019; Grippa *et al.*, 2019; Önenli, 2019; Bernal and Ocampo, 2020; Gökmenoglu *et al.*, 2021). These studies either used scenario-based analysis or monitored a situation.

The aim of this chapter is to review green swan events with the financial and price stability of the economies by focusing on Annex II countries and Turkey. There are two reasons for evaluating Annex II countries as follows: (i) most of them are responsible for global warming; (ii) they are listed by the United Nations (UN) as a leading group for the transition to a low-carbon economy. Turkey, among the 196 countries, has also committed to decreasing greenhouse gases (GHGs) emissions by 21% until 2030 compared to the business-as-usual approach. This implies that Turkey is responsible for both contributing to the fight against climate change and protecting its financial system from green swan events. As the fastest growing economy in the G20, Turkey is counted as one of the world's next big

---

[1]https://www.ngfs.net/en/about-us/governance/origin-and-purpose.

emerging economies by the World Economic Forum. Although the Eleventh Development Plan covering the period 2019–2023 highlights Turkey's main policy aimed at promoting green growth and limiting the rising trend of emissions by planning a series of actions at sectoral and geographical levels, there is no specific emphasis on the relation between climate-related risks and financial (and price) stability. Therefore, it will be important for the Central Bank of Republic of Turkey (CBRT) to study the implications of climate change for the economy and financial system and to adapt its work accordingly. Recently, the COVID-19 pandemic, which is also thought to be one of the green swan events, has shown that monetary policymakers must accurately assess how labor markets, household and business spending, output and prices are affected (Brainard, 2019).

This chapter consists of five sections. The second section describes green swan events in relation to black swan events. The third section summarizes the current situation of the world and Turkey related to climate change by looking at the most recent data. The fourth section is about climate scenario analyses at micro and macro levels. The chapter ends with discussion and conclusion.

2. Green Swan and Black Swan

The Bank for International Settlements (BIS) published a book entitled "The Green Swan" (Bolton *et al.*, 2020a), in which these events were defined as those that occur with the physical consequences of climate change. In other words, the green swan is the economic consequences of the risks arising from climate change or global warming. Therefore, it differs from a black swan in that it is an unpredictable, rare and important event. Goose (2020) questioned whether the next black swan will be green with the COVID-19 pandemic since, the pandemic started after the BIS published their book on green swan. According to some authors, this is a sort of physical consequence of climate change and also a green swan event (Bolton *et al.*, 2020b; Goose, 2020).

In order to understand a green swan, first the concept of a black swan needs to be considered. The phrase "black swans" was first used as a metaphor for both rare and important events in a book written by Taleb (2001, 2007). Black swan events have the following three features (Taleb, 2001, 2007): (i) appearing surprisingly and unexpectedly; (ii) causing extreme impacts; (iii) can only be analyzed after the event has occurred. These features can occur in many forms, such as technological improvements, natural disasters, wars and terrorist events.[2] Besides, the consequences can be positive or negative although the events are unpredictable. Taleb (2007) indicated that these characteristics of the black swans are important for understanding the complex economic systems rather than focusing on regular events.

According to Newbold *et al.* (2013), the black swan events can be considered as outliers or just values that deviate from the data because they are not predictable via a mathematical and/or statistical methodology. On the other side, there is a white swan definition that is more predictable than a black swan. A white swan can be defined as one that is more predictable even though highly certain events also have main features: its impact can easily be estimated and there is a certain explanation after the event occurs. For this reason, the white swan field is more convenient for empirical researches. However, the green swans have similarities and differences with both the black swans and white swans. Primarily, the green swan is concerned with both climate and ecological related risks, and this is why, as Bolton *et al.* (2020b) highlighted, COVID-19 can be counted as a green swan event. According to their paper, the reasons are that the pandemic is connected with biodiversity erosion and raises systemic issues. The concept of green swan events has two additional characteristics to the three-ingredient definition of black swans. First, the events concerning climate-related risks are thought of as nearly certain to occur although there is no idea about when/where/how they will happen. Second, the green swan events are "irreversible and potentially civilizational" and so a systemic response has to be found as a precaution (Bolton *et al.*, 2020b).

The differences in the swans are summarized in Table 1. In general, it is hard to determine which swan is worse than the other, green or black. It is, however, clear that the white swans are more convenient for analysis and improvement in a methodological way. Specifically, the green swans arise at uncertain time and have very huge impacts. Even though they cannot be predicted, this makes green

---

[2]Khan [2019] gives a very good chronolgy for balck swan events.

Table 1. Similarities and differences of swans.

	White Swans	Black Swans	Green Swans
Predictability through	Gaussian, normal distributions	Tail risks, perhaps non-Gaussian, *ex-post* rational explanation after occurrence	Highly likely or certain occurrence but uncertain timing of occurrence and materialization; too complex to fully understand
Main explanation by	Statisticians, economists	Economists, financial analysts and risk managers	Scientists, disagreement with many economists and financial analysts
Impacts	Low or moderate	Massive and direct impact mostly material; possible correction of damages after event (crisis)	Massive and direct impact mostly to human lives (or even civilizational); irreversibility of damages in most cases
Policy recommendations	Risk models are fine (can be marginally improved)	Reconceptualize approach to risk; learn from event to design anti-fragile strategies	Given severity of effects, even without full understanding, need for immediate action and coordination under radical uncertainty

Source: Pereira da Silva (2020).

swans very attractive for scientists. But economists and financial analysts are more interested in this area due to their effects on the financial stability that is discussed in the next section after reviewing the current climate-related situation of the world and Turkey.

2.1. *The world and Turkey: Current situation*

Commitments of countries to limit the increase in global average temperature to well below 2°C require transition to low-carbon economies. Figure 1 shows the plot of changes in global average temperature with time relative to 1880–1900. As seen in Fig. 1, the changes in the global average temperature have an S-shaped pattern till the 1970s and then they steadily increased. Therefore, delays in this transition will harm the target of net zero by 2050. Net zero demands a huge decline in the use of fossil fuels that, as of today, responsible for about 70% of carbon dioxide (CO_2) emissions.

Figure 2 shows time trend of the share of CO_2 emissions sourced by fossil fuel use in the distinction of liquid and solid sources. While the share of liquid fossil fuels in the world's total CO_2 emissions continuously increased in the 1960–1973 period, the contribution of solid fossil fuels decreased. In the 1973–1979 period, there was no significant change in the contribution of both types of fossil fuels to the total CO_2 emissions. Contributions of these two types of fossil fuels to the global CO_2 emissions are approximately same. While the share of liquid fossil fuels has continuously decreased as of the end of the 1970s, the share of solid fossil fuels has generally increased. Although CO_2 emissions from fossil fuels tend to decrease over time,

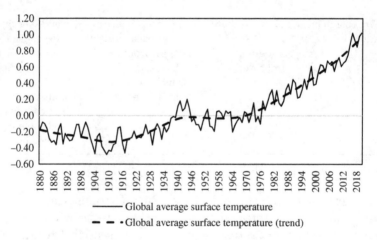

Fig. 1.   Changes in global average temperature relative to 1880–1900.

Note: The y-axis shows changes in the global average temperature relative to 1880–1900 in Fahrenheit. The trend in the changes in global average temperature relative to 1880–1990 was estimated by the authors with the use of the Hodrick–Prescott filter and available upon request.

Source: https://data.giss.nasa.gov/gistemp/ (Accessed Date: 14/05/2021).

its share is still around 70%, meaning that fossil fuel is still the predominant source of energy generation in the world.

Figure 3 exhibits CO_2 emission trends for Turkey. Turkey follows a similar pathway with the world in general. The share of the use of solid fuels in the total CO_2 emissions in Turkey has continuously increased until 1978. These years are the years when import substitution growth policies were implemented and it was aimed to produce

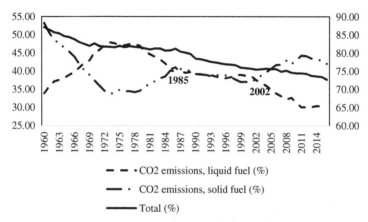

Fig. 2. Global CO_2 emissions sourced by liquid and solid fuel consumption, 1960–2016.

Source: The World Development Indicator.

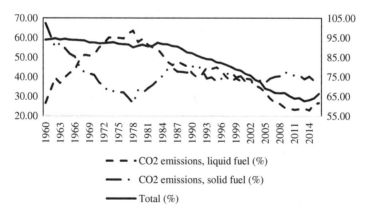

Fig. 3. Turkey: CO_2 emissions sourced by liquid and solid fuel consumption, 1960–2016.

Source: The World Development Indicator.

important industrial products in Turkey. The CO_2 emissions from liquid fuels started to exceed emissions from solid fuels in the early 2000s. It can be said that the increase in the use of natural gas and partially renewable energy and a relative decrease in the use of coal have contributed to this pattern. As a matter of fact, as of 2017, the share of hydraulics in Turkey's installed capacity is 32.01%, natural gas is 27.07%, coal is 21.9% and wind, solar and other renewable energies is 13.6% (Önenli, 2019). Although the share of fossil fuels in CO_2 emissions has a decreasing trend, it shows that it will tend to increase in the next few years. In other words, the use of fossil fuels in energy generation produces at least 65% of the CO_2 emissions in Turkey, which is close to the world's average.

Renewable energy is the main reason for decreasing CO_2 emissions. Renewable energy sources are classified as hydropower, wind, solar and bioenergy to produce energy and/or electricity. Table 2 summarizes electricity generation (GWh) in terms of renewable energy sources based on geographical regions. According to Table 2, the share of hydropower technology in the total renewable energy decreased from 72.2% to 64.8% in the world during 2015–2018. This share is higher in Africa, Asia, Eurasia, Middle East and South America while Central America and Caribbean Islands, Europe, EU 28, North America and Oceania have lower shares than those for the world in 2018. The use of wind for electricity generation increases in the world on average. Eurozone as a whole, North America and Oceania have the highest contribution to electricity generation by using wind technology. Specifically, Turkey has the same trend as the world average. While solar and bio technologies have almost the same contributions to electricity generation in the world on average, countries in the Middle East, EU 28, Oceania and Asia stand out in the use of solar technology for electricity generation. Turkey has a big jump in 2018 than the previous years in using solar energy. The Eurozone has the highest share in bioenergy technologies for generating electricity.

According to the data provided by the International Renewable Energy Agency (IRENA), renewable energy investments are concentrated in East Asia and the Pacific (includes China), which is 32%. The second-largest share of renewable energy investments

Table 2.   Shares of hydropower, wind, solar and bioenergy in total renewable energy for the selected areas and years (%).

Country/Area	Technology	2015	2016	2017	2018
World	Hydropower	72.2	70.6	67.3	64.8
	Wind	15.0	16.2	18.3	19.2
	Solar	4.6	5.5	7.1	8.5
	Bioenergy	8.4	8.2	8.0	7.9
Africa	Hydropower	89.3	86.0	85.3	83.7
	Wind	5.5	7.2	7.7	8.8
	Solar	2.4	3.8	4.8	5.4
	Bioenergy	2.1	2.2	1.9	2.0
Asia	Hydropower	78.6	75.7	71.2	67.0
	Wind	11.2	13.1	15.2	16.6
	Solar	4.4	6.1	8.6	11.1
	Bioenergy	6.2	6.2	6.2	6.2
C. America+ Carib	Hydropower	63.5	61.9	65.7	61.2
	Wind	11.4	11.2	10.0	12.3
	Solar	3.1	4.6	4.9	5.5
	Bioenergy	11.5	12.3	10.8	12.7
Eurasia	Hydropower	94.2	92.5	90.3	87.8
	Wind	4.4	5.4	6.3	6.6
	Solar	0.1	0.4	1.1	2.7
	Bioenergy	0.5	0.6	0.8	0.9
Europe	Hydropower	50.9	51.4	46.1	47.2
	Wind	26.2	25.8	30.4	29.6
	Solar	9.4	9.5	10.1	10.2
	Bioenergy	15.4	15.3	15.4	14.7
EU 28	Hydropower	39.7	40.0	34.1	36.1
	Wind	32.2	31.8	37.1	35.9
	Solar	11.6	11.7	12.3	12.2
	Bioenergy	18.9	18.9	18.9	17.9
Middle East	Hydropower	87.8	84.1	79.2	70.4
	Wind	1.8	2.5	3.0	5.0
	Solar	8.6	12.0	16.4	23.2
	Bioenergy	1.8	1.4	1.4	1.4
N. America	Hydropower	65.8	63.4	62.4	59.7
	Wind	21.9	24.1	24.7	26.1
	Solar	3.7	4.8	6.3	7.4
	Bioenergy	8.1	7.4	6.5	6.6

(*Continued*)

Table 2.   (*Continued*)

Country/Area	Technology	2015	2016	2017	2018
Oceania	Hydropower	55.8	56.3	55.1	52.7
	Wind	19.5	19.1	18.8	20.6
	Solar	7.2	8.4	10.6	12.2
	Bioenergy	6.1	5.8	5.3	5.0
S America	Hydropower	87.7	86.1	84.2	82.5
	Wind	3.7	5.2	6.6	7.6
	Solar	0.2	0.4	0.7	1.4
	Bioenergy	8.4	8.4	8.5	8.6
Turkey	Hydropower	80.3	74.5	66.7	61.3
	Wind	13.9	17.2	20.5	20.4
	Solar	0.2	1.2	3.3	8.0
	Bioenergy	1.5	1.8	2.4	2.7

Source: IRENA (2021), Renewable Capacity Statistics 2021; & IRENA (2020), Renewable Energy Statistics 2020, The International Renewable Energy Agency, Abu Dhabi.

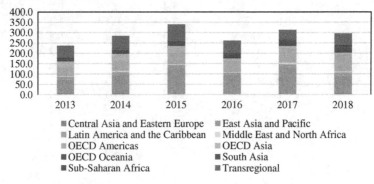

Fig. 4.   Annual financial commitments in renewable energy by region.
Source: IRENA and CPI (2020), Global Landscape of Renewable Energy Finance 2020, International Renewable Energy Agency, Abu Dhabi.

belongs to Western European countries with 19% (see Fig. 4). Sub-Saharan Africa and South Asia have less share in the contribution to renewable energy investments among the other regions. Total renewable energy investment is USD 236.9 billion in 2013 and it became USD 296.4 billion in 2018.

As noted in the Special Report of the International Energy Agency (IEA) (2021), transition to a low-carbon energy economy requires a big expansion in investment and a big shift in what capital is spent on. Annual capital investment in energy is globally around USD 2 trillion on average in the 2015–2020 period. It is expected to increase to USD 5 trillion by 2030 and to USD 4.5 trillion by 2050.[3] This means that investments in renewable energy need to be increased. Therefore, economic stability is a vital ingredient of the transition process, and it deserves to look into the details of the current economic situation of countries — especially for Annex II countries listed by the UN as a leading group for such a transition.

3. Climate-related Green Swans and Economic Stability

Price stability has become important more than ever for fighting against climate change in terms of reducing the CO_2 emissions since the fight requires a transition to a low-carbon economy. As a general increase in the level of prices in economy, which is called inflation, implies that prices do not fluctuate around a certain level but continuously increase instead. On the contrary, a slow decrease in the rate of inflation, called disinflation, also affects price stability that refers to a balance between inflation and disinflation. As an accepted inflation rate of about 2%, price stability is important for the economies because it contributes to the economies' growth and employment prospects in the long term by moderating the variability in output and employment in the short term (Bernanke, 2006). In other words, price stability refers to a low and stable inflation rate that will not be effective in the decision-making processes of economic agents for growth and employment. High inflation can lead to a spiral of increasing prices and limits purchasing power of the money by making prices go up. Increasing prices decrease sales of businesses since demand falls and this leads to reduction or freezing of wages or decrease in the number of employees. Decreasing

---

[3]*Source*: https://iea.blob.core.windows.net/assets/063ae08a-7114-4b58-a34e-39 db2112d0a2/NetZeroby2050-ARoadmapfortheGlobalEnergySector.pdf (Accessed Date: 20/05/2021).

production will result in smaller or negative economic growth and deflation also harms the economy in a similar way. Increases in expectations of people that the prices will go down bring about decreases in sales and therefore possible increases in unemployment. This will lead to slowing down in the economy as consumers and businesses cut back on spending and investing. In brief, the real value of money decreases and people do not rely on their own currencies as a measure of value when prices are continuously high and disinflation causes uncertainty.

As explained above, climate change affects the economies both in a direct and indirect way. Its direct effects rise from environmental shifts while indirect effects come with the transition to a low-carbon economy. Villeroy de Galhau (2019), Governor of Banque de France, discusses the impacts of climate change on pricing in two ways. First, more regular adverse weather conditions potentially create more volatile agricultural and food prices. Since energy generation sources will be replaced by renewables, extraction and supply difficulties could lead to sharp price adjustments. Therefore, this can intensify medium term inflationary pressures through repercussions of food and energy prices on production costs. Second, extreme weather conditions will affect infrastructure, buildings, the health of employees and productivity, resulting in instability of prices. For this reason, the existing economic policies might not be as effective as in the past and present to maintain price stability under climate change (Coeuré, 2018).

As noted above, price stability and sustainable economic growth are the main goals of monetary policy. Since the fight against climate change requires financing and central banks monitoring the system, financial stability also needs to be taken into account in this campaign. Even though defining financial stability is not as easy as defining price stability because of complexity of financial systems, it is generally defined as the resilience of the economy against unexpected situations that may disturb the balances of the financial system. The European Central Bank (ECB) defines financial stability as a condition in which the financial system — which comprises financial intermediaries, markets and market infrastructures — is capable of withstanding shocks and the unravelling of financial imbalances.[4]

---

[4]ECB also has discussed alternatives to this definition of price stability. Details of this discussion can be found in Bremus *et al.* (2020).

Economic agents and decision makers will be confident about the near future of the economy in a financially stable economy. However, one might think that climate-change-related financial risks can be more severe than the usual ones because of transition risks. These types of risks come from the energy and industrial transition (TCFD, 2017; Chenet, 2019). This transition will affect productive enterprises of today since their production mainly depends on employing energy sources that are carbon-accelerated and the technology they use.

The TCDF (TCFD, 2017)[5] classifies transitional risks into five groups: policy and legal risks, technology risk, market risk and reputation risk. Policy actions are either attempting to constrain actions that contribute to the adverse effects of climate change or seeking to promote adaptation to climate change. Risks come from timing of policy actions since Earth does not have even a minute to lose in this fight. It is possible to face with litigation risk because the value of loss and damage arising from climate change will grow during the adaptation process. The timing of technology development and deployment is a key uncertainty in assessing technology risk. Shifts in supply and demand for certain commodities, products and services will create market risk. Enterprises will face a loss of their reputations if they do not change their perceptions in the eyes of customers and community. However, risks are taken not only by financial institutions but also their clients. Climate change for banks was essentially limited to carbon markets and project finance, while the responsibility over GHG emissions was fully left to the banks' clients' decisions until the 2010s (Chenet, 2019). The role of financial institutions — banks and central banks — has been accelerated since COP21 in 2015.

Financial transition risks can be described as the risks coming from the energy and industrial transition in the context of climate

---

[5] "Communiqué from the G20 Finance Ministers and Central Bank Governors Meeting in Washington, D.C. April 16–17, 2015," April 2015.

TCFD, fighting against climate change is leading to new boards or institutions to watch or to force. Given such concerns and the potential impact on financial intermediaries and investors, the G20 Finance Ministers and Central Bank Governors asked the Financial Stability Board to review how the financial sector can take account of climate-related issues.

change. Financing such a radical transformation of the economy will require massive redirection of financial flows, which will clearly not be possible without substantial contribution of capital markets (Louche *et al.*, 2019). A detailed answer to the question of why financial stability matters in combating climate change are given with two main channels (IMF, 2020). The first channel relates a climatic hazard to the total life, including people, livelihoods, natural resources and infrastructure as well as the socio-economic and cultural environment. Households, non-financial firms and the government sector will be affected if a disaster occurs, and this will lead to both loss of physical and human capital. Financial institutions could also be exposed to operational risk or liquidity risk, and among them, insurance companies might be asked to bear all burdens. The second one is built upon invertors' beliefs about the transition risks. The assets might be stranded which is counted as a new financial risk. Under such conditions, the pricing mechanism might not work in a desired or expected way and lead to financial instability.

Grippa *et al.* (2019) visualized these physical and transition risks, as shown in Fig. 5. In brief, physical risk is related to extreme weather events, such as hurricanes and rising sea levels, and their long-term effects. Transition risk refers to financial implications of policy changes, important technological improvements, market preferences, changes in social norms, etc. The former affects the economic

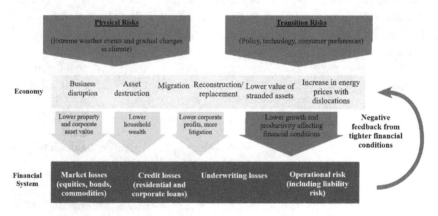

Fig. 5.   Physical and transition risks.
Source: Grippa *et al.* (2019).

structure through business disruption, asset destruction, migration and reconstruction replacement, while the latter leads to lower-valued stranded assets and repricing in energy with dislocations. Market, credit and underwriting losses with operational risks will direct the financial system that must be robust in combating these effects of climate-related risks.

3.1. *What does the recent past tell about the near future? Let the data speak*

Combating climate change has important milestones including conventions, a protocol and a main agreement. The United Nations Framework Convention on Climate Change (UNFCCC) entered into force on March 21, 1994. Today, it has near-universal membership. The 197 countries that have ratified the Convention are called Parties to the Convention. Preventing "dangerous" human interference with the climate system is the ultimate aim of the UNFCCC. The Kyoto Protocol — adopted on December 11, 1997, and entered into force on February 16, 2005 — is based on the principles and provisions of the Convention and follows its annex-based structure. The protocol only binds developed countries and places a heavier burden on them under the principle of "common but differentiated responsibility and respective capabilities" because it recognizes that they are largely responsible for the current high levels of GHG emissions in the atmosphere.

Annex I Parties include the industrialized countries that were members of the Organization for Economic Co-operation and Development (OECD) in 1992 plus countries with economies in transition (the EIT Parties), including the Russian Federation, the Baltic states and several Central and Eastern European states. Annex II Parties consist of the OECD members of Annex I but not the EIT Parties. They are required to provide financial resources to enable developing countries to undertake emissions reduction activities under the Convention and to help them adapt to adverse effects of climate change. In addition, they have to "take all practicable steps" to promote the development and transfer of environmentally friendly technologies to EIT Parties and developing countries.[6]

---

[6]https://unfccc.int/parties-observers.

The Paris Agreement supplies the countries that need financial, technical and capacity-building support. Therefore, it is better to look at their data related to price and financial stabilities of the Annex II countries. Turkey's claim of contributing much better to combating climate change by being a non-Annex-I country was accepted in December 2018 and since then, Turkey is not listed as a member of the Annex II Parties.

Table 3 lists the inflation rates in 35 Annex II countries in the periods of 1996–2004 and 2005–2020, which are the periods before and after the Kyoto Protocol that entered into force in 2005, respectively.[7] A panel data of Annex II countries is used to calculate the summary statistics of the inflation rates as well as financial stability indicators. For inflation rates, the overall and within standard deviations are calculated for over 306 country–years of data in 1996–2004 period and 544 in 2005–2020 period. The average inflation rate is about 12% and the variation in inflation rates within a country over time is approximately two times higher than the variation across countries. In the post-Kyoto protocol period, even though both the average inflation rate and the variations within a country over time and across countries sharply decreased, the variation across countries is one-and-a-half times higher than the variation in a country over time. This means that inflation rates have different patterns among the Annex II countries in both periods. Although price stability has the same meaning across countries, their methods for maintaining it can differ among them. This country group also includes the EU countries, and the European Central Bank is involved to maintain stability in prices, i.e. to safeguard the value of the euro, in the Eurozone. Thus, the inflation rate was 2.4% in the 1996–2004 period and 1.7% in the 2005–2020 period. It can be said that the Eurozone has a very significant contribution to price stability in Annex II countries.

Table 4 is given to review Turkey's inflation adventure over the period. As a dynamic concept, price stability shows the ability of the system to return itself to a targeted equilibrium after temporary shocks. Turkey followed the money supply anchor and/or

---

[7]It should be noted that each country or country groups have their own composition and survey methodology for calculating consumer price indices. Summary statistics given in this chapter should be taken as broader comparisons among the countries.

Table 3. Annex II countries: Summary statistics of yearly financial stability indicators and inflation rates.

Variable		Mean	S.D.	Min.	Max.	Obs.	CV
				1996–2004			
Inflation	Overall	11.97	64.77	−1.13	1058.37	306	5.41
	Between		27.79	−0.04	137.02		
	Within		58.68	−122.71	933.32		
Bank *z*-score	Overall	11.29	6.42	0.49	34.78	306	0.57
	Between		5.85	0.77	24.23		
	Within		2.81	2.19	27.03		
Bank nonperforming loans to gross loans (%)	Overall	4.74	5.57	0.20	31.60	206	1.18
	Between		4.48	0.44	16.08		
	Within		3.46	−6.31	22.69		
Stock price volatility	Overall	7.43	3.08	2.00	19.00	199	0.42
	Between		3.11	3.26	17.60		
	Within		1.27	2.51	11.43		
				2005–2020*			
Inflation	Overall	2.62	4.28	−4.48	59.22	544	1.63
	Between		2.76	0.29	16.09		
	Within		3.31	−8.59	45.75		
Bank *z*-score	Overall	12.13	7.88	0.02	47.57	441	0.65
	Between		7.48	1.71	33.47		
	Within		2.77	2.22	26.23		
Bank nonperforming loans to gross loans (%)	Overall	5.93	7.45	0.10	48.68	409	1.26
	Between		5.68	0.43	26.24		
	Within		5.14	−16.71	31.58		
Stock price volatility	Overall	7.80	3.13	2.70	21.10	401	0.40
	Between		2.66	4.64	15.32		
	Within		1.60	0.11	14.13		

Note: *Data for financial stability indicators cover the period 1996–2017.
The coefficient of variation (CV) is used to compare the degree of variation from one variable to the other.
Each variable's (x_{it}) variation according to country and time is also decomposed into two: A between ($\bar{x}_i$) and a within ($x_{it} - \bar{x}_i + \bar{\bar{x}}_i$), where $\bar{\bar{x}}_i$ is the global mean and it is added back in to make the results comparable.
Source: Financial stability indicators are retrieved from the latest version of Global Financial Development Database of the World Bank. Inflation rates are retrieved from the World Development Indicators of the World Bank except 2020 data that come from International Financial Statistics of International Monetary Fund.

Table 4.   Turkey: Summary statistics of yearly financial stability indicators and inflation rates.

	Obs	Mean	S.D.	Min.	Max.	CV
Variable			1996–2020			
Inflation	25	26.13	27.56	6.25	85.67	1.05
Bank z-score	22	7.91	2.24	0.42	11.48	0.28
Bank nonperforming loans to gross loans (%)	20	6.51	6.25	2.58	29.30	0.96
Stock price volatility	20	11.09	2.44	5.20	15.00	0.22
			1996–2004			
Inflation	9	55.56	27.22	8.60	85.67	0.49
Bank z-score	9	6.60	2.99	0.42	11.48	0.45
Bank nonperforming loans to gross loans (%)	7	12.34	7.83	6.50	29.30	0.63
Stock price volatility	7	9.73	3.76	5.20	15.00	0.39
			2005–2020			
Inflation	15	9.67	2.91	6.25	16.33	0.30
Bank z-score	12	8.71	0.73	7.96	9.73	0.08
Bank nonperforming loans to gross loans (%)	12	3.23	0.68	2.58	5.00	0.21
Stock price volatility	12	11.70	0.71	10.72	12.80	0.06

Notes: *Data for financial stability indicators cover the period 1996–2017.
The coefficient of variation (CV) is used to compare the degree of variation from one variable to the other.
Each variable's (x_{it}) variation according to country and time is also decomposed into two: A between $(\bar{x}_i)$ and a within $(x_{it} - \bar{x}_i + \bar{\bar{x}}_i)$, where $\bar{\bar{x}}_i$ is the global mean and it is added back in to make the results comparable.

the exchange rate anchor to decrease her higher inflation rates before the 2001 economic crisis. With the 2001 Transition to a Strong Economy Program, Turkey's strategy fighting against inflation is the inflation targeting policy and it officially started to implement in 2006, which was following implicitly after the 2001 crisis. The first five years of the inflation targeting period can be characterized as conventional inflation targeting, while the period after 2011 can be characterized as being a more flexible regime which integrated financial stability targets into the inflation targeting (Kara, 2013). This policy helped to reach one-digit levels; however, the inflation rate was 8.72% on average in the 2004–2006 period though the targeted value was 5%. The average inflation rate was recorded as 9.6% in the 2005–2020 period, and since 2011, yearly inflation rates of Turkey are above 11% although inflation rates vary less in the post-Kyoto protocol period. The coefficient of variation is less than half of its value in the pre-Kyoto period. These numbers show that the price stability of the Turkish economy needs time to be established when compared to Annex II countries.

The COVID-19 pandemic still affects humanity even though there is a light at the end of the tunnel due to several vaccine options. Countries have been following restrictions ranging from light to severe to protect their population, healthcare system and economy. The inflation rate of the Eurozone is 0.9% in January 2021 compared to January 2020. France, Germany, the United Kingdom and the United States (US) recorded inflation rates in the same period as 0.6, 0.7, 0.1 and 1.4%, respectively. China and Japan reported negative inflation rates, while India, the Russian Federation and Brazil — all emerging markets — reached 3.2, 5.2 and 4.6% inflation rates in January 2021. The inflation rate of the US was declared as 4.2% in April 2021 compared to April 2020, which is the highest inflation rate observed in the US since 2008. It is more likely that the COVID-19 pandemic will impact inflation rates across many countries, and this might hurt combating climate change now and in the near future.[8]

Financial stability performances of Annex II countries during the period 1996–2020 are also assessed. Three indicators are summarized in Tables 3 and 4 to evaluate the financial stability performances of the countries. The first one is "Bank z-score" which helps to capture the probability of default of a country's banking system. The buffer factors of a country's banking system (capitalization and returns) and the volatility of their returns can be compared via this score.[9]

---

[8]These numbers were collected from the European Central Bank and Bureau of Labor Statistics of the US Department of Labor.

[9]https://www.worldbank.org/en/publication/gfdr/gfdr-2016/background/finan cial-stability.

The World Bank estimates this score as (ROA + (equity/assets))/ sd(ROA), where sd(ROA) is the standard deviation of return of assets (ROA). A higher z-score implies a lower probability of insolvency for the countries that is adapted from firm-level data by the World Bank. The second indicator is obtained from the ratio between the defaulting loans (payments of interest and principal past due by 90 days or more) and total gross loans. This ratio is called as non-performing loans (NPL), and a higher NPL indicates that countries are holding riskier assets.[10] It should be noted that these data are not definitely comparable for countries since the differences in national accounting, taxation and supervisory regimes. However, it is worth using this indicator for Annex II Parties that have a larger proportion of the European Union countries.[11] The third one which is stock price volatility calculated as the average of the 360-day volatility of the national stock market index. It is more difficult to forecast the returns of stocks than the variability of these returns (Hamilton and Lin, 1996). Stock market volatility measures fluctuations in the stock market's overall value. Higher volatility implies that the stock market being considered shows unstable characteristics which means dramatic changes in the prices of stocks over a short time period in either direction.

According to Table 3 given for Annex II countries, while the average bank z-score was 11.29 in the 1996–2004 period, this score increased by 0.84 points and became 12.13 in the Kyoto Protocol period that covers the 2005–2017 period. In the first sub-period, the variability of this score across countries (6.42) is about one-fold higher than the variability of a country's over time (5.85). In the period 2005–2020, these variations are almost equal meaning that

---

[10]According the data definition given by the World Bank, capital and reserves include funds contributed by owners, retained earnings, general and special reserves, provisions, and valuation adjustments. Capital includes tier 1 capital (paid-up shares and common stock), which is a common feature in all countries' banking systems, and total regulatory capital, which includes several specified types of subordinated debt instruments that need not be repaid if the funds are required to maintain minimum capital levels (these comprise tier 2 and tier 3 capital). Total assets include all nonfinancial and financial assets (https://datacatalog.worldbank.org/bank-capital-assets-ratio-1).

[11] https://www.worldbank.org/en/publication/gfdr/gfdr-2016/background/financial-stability.

if we were to draw two countries randomly from the data, the difference in bank z-score is expected to be nearly equal to the difference for the same country in two randomly selected years. This implies that Annex II countries show a more financially stable structure during the Kyoto Protocol period than that during the previous period. Even though a strict comparison should be avoided, it can be said that countries' bank nonperforming loans to total gross loans slightly increased in the Kyoto Protocol period. The variability of this ratio across countries in 1996–2004 is slightly higher than those in the 2005–2017 period, implying that banks in these economies might have riskier assets. Though stock markets of Annex II countries shared almost the same volatility on average in these two periods, the variability of stock volatility across Annex II Parties is significantly greater in the Kyoto Protocol period. This can be interpreted as that the stocks are riskier investments in the second sub-period.

Summary statistics for financial stability indicators were calculated till 2017. The indices calculated by IMF show that financial conditions of the US economy have tightened but remain accommodative. The early-2019 rebound in markets helped attenuate the tightening in financial conditions in the Eurozone. However, in China, financial conditions remain almost stable as market pressures have been balanced by policy easing. Financial conditions in other emerging markets have been stable in early 2019, after tightening in 2018, because of higher external borrowing costs.[12]

The financial health of the financial institutions in Turkey is detailed in Table 4. Even though the average bank z-score of Turkey is approximately higher than one point in the 2005–2017 period, the variability of this score is higher in 1996–2004. It can be said that even if the banking sector of the Turkish economy has a higher bank z-score, i.e. financially less risky, its stability might be vulnerable due to higher variation. The NPL ratio is 6.5 on average in 1996–2004 and it varies between 2.56 and 29.3. The average NLP ratio is higher in the Kyoto Protocol period with changes between 2.56 and 5.0. This summary statistics imply that the Turkish banking system has fewer risky assets in 2005–2017 than in the previous period.

---

[12]https://www.imf.org/-/media/Files/Publications/GFSR/2019/April/English/ch1data/charts/Figure1-2.ashx (Accessed Date: 20/05/2021).

Average stock volatility is around 11 in both sub-periods. However, the stock market in Turkey is more volatile in 1996–2004 period. The Turkish economy has been experiencing continuously increasing exchange rates with considerably high volatility since early 2017. Correlation between yearly inflation rate and exchange rate volatility is significant (about 51%) in the 2005–2020 period.[13] This means that higher volatility in exchange rates leads to higher inflation rates resulting in instability in prices.

As data show, the Annex II countries, in general, are closer to establishing stability in their prices and financial stability as well. A predictable path in transition to a low-carbon economy as well as a predictable economic development will help all countries in their fight against climate change. However, this relatively predictable path has been affected by the COVID-19 pandemic. According to a report prepared by the Department of Economic and Social Affairs Economic Analysis of the UN,[14] world output shrank by 4.3% in 2020. In almost all regions in the world, only East Asian countries recorded positive economic growth in 2020. The pandemic hit the developed countries (with a 5.6% contraction) the hardest while its effect on developing countries is relatively mild (with a 2.5% contraction). Among developing countries, East Asia performed much better than all other developing regions in 2020.

The focal point of economic policy measures is providing liquidity and income support. The aim of liquidity support is to help businesses stay afloat, while the aim of income support is to help vulnerable households (OECD, 2020). However, the UN report says that an estimated 131 million people have been pushed into poverty because of job and income losses in 2020. In brief, this pandemic is unique among the others because of its effects on both supply side (factory and business shutdowns and disruptions in supply chains) and demand side (unemployment and changes in consumers' consumption

---

[13] Authors' calculations using consumer price index (2003 = 100) retrieved from TURKSTAT and nominal exchange rates (USD/TL) retrieved from CBRT. Volatility is simply monthly standard deviations of daily exchange rates.

[14] https://www.un.org/development/desa/dpad/publication/world-economic-situation-and-prospects-february-2021-briefing-no-146/ (Accessed Date: 05/19/2021).

habits) of the economies, and recovery will take longer time to get back to the pre-pandemic trends.[15]

Global capital markets and financial institutions of the economies were also affected by the COVID-19 pandemic, especially in the first months (Ashraf, 2020; Mazur *et al.*, 2020; Narayan *et al.*, 2020; Topcu and Gulal, 2020; Hong *et al.*, 2021). The empirical results obtained in these studies show that there were negative market reactions in the first months of the pandemic and these reactions vary over time depending on the stage of outbreak. Studies (Özkan, 2020; Ünal, 2020; Contuk, 2021) showed that the stock market of the Turkish economy also reacted negatively in the first months of the pandemic and its negative effects, especially on the volatility of the stock market, attenuated. According to the KPMG report (2020), the banking sector is one of the most likely to be affected, especially valuation and profitability. In this way, banking stocks have been impacted during COVID-19 too. This report also notes that most banks faced a price slump in mid-March in the period from December 1, 2019 to April 30, 2020. Also, European banks were adversely affected and the Euro STOXX banks index declined incredibly by 40.18%, followed by STOXX North America 600 banks index (31.23%) and STOXX Asia/Pacific 600 banks index (26.09%). Global Financial Stability Report of the International Monetary Fund highlights that some emerging and frontier market economies face financing challenges leading to financial instability (IMF, 2020).

As reports and academic articles imply that controlling stabilities in the economies will require more effort, this could possibly influence their transition to a low-carbon economy in the near future. According to A Roadmap for the Global Energy Sector Report of the IEA in May 2021,[16] the number of countries that committed to achieving net-zero emissions increased over the last year and it reached almost 70% of global emissions of CO_2. However, these pledges have not been supported by these countries by the near-term policies and

---

[15]https://www.worldbank.org/en/publication/global-economic-prospects (Accessed Date: 05/19/2021).

[16]https://iea.blob.core.windows.net/assets/4482cac7-edd6-4c03-b6a2-8e79792d1 6d9/NetZeroby2050-ARoadmapfortheGlobalEnergySector.pdf (Accessed Date: 05/19/2021).

measures yet. One might expect that the reason behind not exhibiting a clear pathway is the continuing effects of the COVID-19 pandemic shock that have taken a key part in their economic recovery agenda. As explained in the same report, global emissions decreased in 2020 because of the COVID-19 pandemic, although it is already increasing because of economic structures trying to recover. Therefore, delays in acting to reverse that trend will threaten the efforts to reach zero-carbon target by 2050.

4.   Climate Scenario Analysis

Possible physical risks arising from more frequent and severe weather conditions and transition risks arising from low-carbon economy targets need to be evaluated together with the price and financial stabilities. Scenario analysis is a useful tool to understand the possible effects of such risks on the near and/or far future of the economies. It is different from forecasting and projections. Forecasting is generally used for short-term needs and they generally are based on a model or models. Time-series econometric techniques have rigid assumptions in the estimation process and even though projections cover longer periods, they are generally a combination of forecasts and some assumptions for today and the near future. Therefore, a scenario analysis is a hypothetical building block.

Humankind has been experiencing weather events, such as hurricane Katrina and California wildfires, resulting in billion-dollar damages. It is certain that climate change is the reason for such events, but it is not certain when and where they will happen. Therefore, within the climate change context, scenario analysis compares the different possible futures (Chenet, 2019). Such scenario analyses are related to not only possible scenarios about green swan events but also economic conditions. Under different circumstances, one can define possible scenarios for the macroeconomic variables, such as the real gross domestic product, economic growth, inflation and stock market volatility. However, it is difficult to define physical and transition risks. As discussed in the section on economics of climate change, financial institutions and central banks are already aware of implications of financial stability on climate change. For example, financial institutions have incorporated climate risk scenarios into their stress tests that show the health of a financial system. The IMF have addressed physical risks from natural disasters in its stress tests in order to capture financial effects of more frequent and larger natural disasters. In other words, stress tests can be used for industrial companies in their transition to a low-carbon economy. Thus, banks and insurance companies provide such services for their customers who want to switch their businesses into a low-carbon business.

The TCDF determines the basics of using scenario analysis in explaining climate-related risks and opportunities. Climate-related scenarios provide exploration and understanding of the effects of the physical and transition risks on businesses over time in the climate change context. In addition, scenario analysis also helps financial institutions to test the resiliency of their portfolios against climate change.

The IEA and Intergovernmental Panel on Climate Change (IPCC) have developed scenarios for physical and transition risks toward assessing future vulnerability to climate change which are used by scientists and policy analysts. These sorts of scenarios are called transition scenarios and physical climate scenarios. Transition scenarios concern with the conclusions of various policy outcomes upon the energy and economic pathways for achieving the aimed temperature increases with some probability. These scenarios have different assumptions about the different factors to have a climate-friendly economy, such as the timing of policy changes, changes in the energy mix. The 2°C scenario of the IEA is about the pathways for achieving the limit of global warming at or below 2°C. Other 2°C Scenarios are defined by IRENA with renewable energy map (REmap), Greenpeace with advanced energy revolution (R), IPCC with RCP 2.6 and deep decarbonization pathways project. For example, the IEA World Energy Outlook 450's transition scenario defines temperature impact range as 2°C, with a likelihood of around 50%. It assumes that the world population to grow by 0.9% per year, from 7 billion in mid-2012 to 9 billion in 2040 with world GDP to grow at a rate of 3.4% in 2012–2040.[17]

Physical climate scenarios focus on a range of atmospheric GHG concentration. The RCP scenarios concern with the amount of concentration in the atmosphere and the resulting changes in global

---

[17]Details can be found on web pages of the International Energy Agency and the Financial Stability Board.

temperatures (and other variables such as precipitation) at various future points (i.e. out to 2035, mid-century [2046–2065] and end of the century [2081–2100]). Services of such agencies and other institutions, such as the World Resources Institute and US Environmental Protection Agency, include several tools that are available to organizations to support their assessments of transition and physical climate impacts and risks at global, regional, national and local levels.

A macro-scenario analysis is also challenging in the context of climate change. It can be considered in two ways: instabilities due to possible climate change events and instabilities in an economy face with climate change mitigation and adaptation policies. Therefore, scenarios should include details about ecological, regional, financial, distributional and social basis. For example, Victor (2012) presented several macroeconomic scenarios for the Canadian economy to understand the alternative paths in this context. The author defined four scenarios: business as usual, a low-/no-growth scenario, selective growth and degrowth scenarios. The main question of this study is how much breathing room the high-consumption economies need to leave to less-developed and developing countries — that will benefit from high economic growth — by developed countries that are responsible for increasing global temperature to meet ambitious greenhouse gas reduction targets.

A scenario analysis was conducted by Önenli (2019) to clarify whether renewable energy can be considered as a solution to meet the emission reduction targets and satisfy the growing electricity demand of Turkey. The author designed the following four scenarios to put forth impacts of different energy generation mix on GHGs emissions: (i) business as usual that keeps the existing generation mix; (ii) maximize local which is using local resources both renewables and local coal which are lignite and hard coal; (iii) generation mix with nuclear which accommodates 4800 MW fully operational nuclear power plant by 2023; (iv) minimize GHGs emissions which is considered as the ideal scenario that carries onward in terms of achieving not only emissions targets but reaching zero GHGs emissions. Based on the business-as-usual scenario, current overcapacity by 2026 could create barriers for the transition of low-carbon generation and today's generation mix will result in a 138% increase in the GHGs emissions. Under maximize-local scenario, GHGs emissions will peak in 2030 with increasing use of renewables. According to the third scenario, the lowest GHGs emissions is possible because of generation mix with nuclear that allows to increase renewable capacity gradually and minimize the share of fossil fuels.

Telli *et al.* (2008) used a general equilibrium model for Turkey over the period 2006–2020 to observe the economic impacts of the planned policy scenarios of compliance with the Kyoto Protocol. The results suggest that the responsibility of emission control targets and the presumed abatement costs could be rather high, and additionally, the expanded abatement investments have to be financed from limited domestic resources.

In this chapter, we reviewed the literature putting effort to understand how climate change affects the financial stabilities of the economies with price stability. Different scenarios are usually used to picture the future with the net-zero policy to see under which circumstances the net-zero target will be reached. These reviews suggest that these types of analysis can be extended to put some assumptions on financial and price stability and climate-related risks within the framework of environmental economics.

5. Discussion and Conclusion

The main motivation of this research is to draw attention to the green swan events and its economic threats. When considering the effects of the COVID-19 pandemic process upon the societies and economies of countries, this issue becomes more important to examine for taking both the ecological and financial precautions and the future green swan events more seriously. From this point of view, the current situation of the world and Turkey are reviewed with data and interaction between climate-risks and financial and price stabilities is explained. Based on the main results, the CO_2 emissions have been decreasing while the global surface temperature has been increasing considering the entire world. Also, the Annex II countries show a financially stable structure during the Kyoto Protocol period than the previous. However, Turkey seems financially less risky although it has higher variation.

The ever-growing literature on green swan events, climate-related risks and emerging new institutions draw attention to the need to ensure financial stability and price stability in combating climate change. Nearly 200 countries have committed to decrease total GHG emissions by 2050 under the leadership of the UN. The World Bank classifies the countries, based on their gross national income per capita, into four categories: low, lower-middle, upper-middle and high-income. The number of low-income economies is 29 while there are 50 lower-middle-income economies in 2020. The national income per capita of these 59 countries is less than USD 4,000. Fifty-six economies are classified under upper-middle-income economies having national income per capita between USD 4,046 and USD 12,535. There are 83 high-income economies with the national income per capita more than USD 12,535. This shows that even though there is a strong commitment to fighting against climate change and its effects on economies, expecting the transition of countries to a low- or zero-carbon economy to be regular and relatively balanced, even if the COVID-19 epidemic did not exist, is not compatible with the realities of the world. Regardless of the classification criterion, there are countries that will find it difficult to keep up with this change, and the UN assigns important roles and responsibilities to Annex II countries for this purpose. Some of these countries may wish to continue to use their available resources — positive or negative contribution to climate change — to industrialize. However, they may also turn to renewable energies while using existing resources, but all of these are activities that require both time and funding. Therefore, countries need to initiate a strict structural change in terms of implementation, with a definite framework for transition to a low-carbon economy and a pathway to follow. Although an economically stable environment will certainly make this transition less painful, it is not enough by itself. While preparing emergency action plans against COVID-19-like or completely different outbreaks in the future, countries from the least developed ones to the most developed need to urge policies to eliminate the preventable effects of possible epidemics to combat climate change.

As supply creates its own demand, combating climate change created organizations such as the FSB, TCFD and the Central Banks and Supervisors of NGFS. However, more research and analyses are needed to include these risks into financial stability, including improvements in data and models, as noted by Brunetti *et al.* (2021) and the Central Bank of the US. Finally, it should be noted that the Central Bank of the Republic of Turkey discussed financial risks stemming from climate change and environmental finance for the first time in its 32nd Financial Stability Report.

References

Ashraf, B.N. (2020). Stock market's reaction to COVID-19: Cases or fatalities? *Research in International Business and Finance*, **54**, 1–7.

Bernal, J. and Ocampo, J.A. (2020). *Climate Change: Policies to Manage Its Macroeconomic and Financial Effects.* (s.l.: UNDP).

Bernanke, B.S. (2006). Semiannual Monetary Policy Report to the Congress.

Bolton, P. *et al.* (2020a). *The Green Swan: Central Banking and Financial Stability in the Age of Climate Change.* (France: Bank for International Settlements).

Bolton, P. *et al.* (2020b). "Green Swans": Central banks in the age of climate-related risk. *Bulletin de la Banque de France*, **229**(8): 1–15.

Bozkurt, C. and Akan, Y. (2014). Economic growth, CO_2 emissions and energy consumption: The Turkish case. *International Journal of Energy Economics and Policy*, 4(3): 484–494.

Brainard, L. (2019). Why Climate Change Matters for Monetary Policy and Financial Stability. [Online] Available at: https://www.federalres erve.gov/newsevents/speech/brainard20191108a.htm [Accessed May 2020].

Bremus, F., Dany-Knedlik, G., and Schlaak, T. (2020). Price stability and climate risks: Sensible measures for the European Central Bank. *DIW Weekly Report*, **10**(14), pp. 206–2013.

Brunaker, F. & Nordqvist, A. (2013). *A Performance Evaluation of Black Swan Investments.* (Sweden: University of Gothenburg).

Campiglio, E. *et al.* (2018). Climate change challenges for central banks and financial regulators. *Nature Climate Change*, **8**: 462–468.

Carney, M. (2015). Breaking the Tragedy of the Horizon–Climate Change and Financial Stability. [Online] Available at: https://www.bis.org/ review/r151009a.pdf[Accessed May 2020].

Chenet, H. (2019). *Climate Change and Financial Risk.* (s.l.: s.n).

Coeuré, B. (2018). *Monetary Policy and Climate Change.* (Berlin, Germany: Deutschebank Bundesbank).

Directorate General, Treasury, Banque de France, and ACPR (2017). Assessing Climate Change-Related Risks in the Banking Sector. *Directorate General of the Treasury.* Accessed May 2020. https:// www.tresor.economie.gouv.fr/Ressources/File/433465.

Gocen, H., Akturk, H. and Orhan, M. (2015). Black Swan Sticking out in Turkish Banking. *Euroasian Journal of Business and Economics*, **8**(15): 1–19.

Goose, S. (2020). James Hambro&Partners. [Online] Available at: https://www.jameshambro.com/insight/comment/will-the-next-black-swan-be-green/ [Accessed 15 April 2021].

Gökmenoglu, K.K. and Taspinar, N. (2016). The relationship between CO_2 emissions, energy consumption, economic growth and FDI: The case of Turkey. *The Journal of International Trade & Economic Development*, **25**(5): 706–723.

Gökmenoglu, K.K., Taspinar, N. and Rahman, M.M. (2021). Military expenditure, financial development and environmental degradation in Turkey: A comparison of CO_2 emissions and ecological footprint. *The International Journal of Finance and Economics*, **26**: 986–997.

Grippa, P., Schmittmann, J. and Suntheim, F. (2019). Climate change and financial risk. *Finance & Development*, 26–29.

Halicioglu, F. (2009). An econometric study of CO_2 emissions, energy consumption, income and foreign trade in Turkey. *Energy Policy*, **37**: 1156–1164.

Hamilton, J.D. and Lin, G. (1996). Stock market volatility and the business cycle. *Journal of Applied Econometrics — Special Issue: Econometric Forecasting*, **11**(5): 573–593.

Higgins, D.M. (2014). Fires, floods and financial meltdowns: Black swan events and property asset management. *Property Management*, **32**(3): 241–255.

IMF. (2020). Global Financial Stability Report: Bridge to Recovery. International Monetary Fund. Accessed May 2020. https://www.imf.org/en/Publications/GFSR.

International Energy Agency, I. (2021). *Net Zero by 2050: A Roadmap for the Global Energy Sector*. (s.l.: IEA Publications).

Kara, Hakan. (2013). Monetary policy after the global crisis. *Atlantic Economic Journal*, **41**(1): 51–74.

Khan, F. (2019). *Data Driven Investor*. [Online] Available at: https://www.datadriveninvestor.com/2019/01/18/9-black-swan-events-that-changed-the-financial-world/. [Accessed 15 April 2021].

KPMG (2020). *Standing Firm on Shifting Sands in Global Banking M&A Outlook H2*. (s.l.: KPMG).

Lleo, S. and Ziemba, W.T. (2015). The Swiss black swan bad scenario: Is Switzerland another casualty of the Eurozone crisis? *International Journal of Financial Stability*, **3**(3): 351–380.

Louche, C., Busch, T., Crifo, P. and Marcus, A. (2019). Financial markets and the transition to a low-carbon economy: Challenging the Dominant Logistics. *Organization & Environment*, **32**(1): 3–17.

Mazur, Mieszko, Man Dang, and Miguel Vega. (2020). COVID-19 and the march 2020 stock market crash. Evidence from S&P1500. *Finance Research Letters*, **38**. doi:10.1016/j.frl.2020.101690.

Nasreen, S., Anwar, S. and Ozturk, I. (2017). Financial stability, energy consumption and environmental quality: Evidence from South Asian economies. *Renewable and Sustainable Energy Reviews*, **67**: 1105–1122.

Newbold, P., Carlson, W.L. and Thorne, B. (2013). Statistics for Business and Economics. 8th edn. (Harlow: Pearson).

OECD (2020). Tax and fiscal policy in response to the Coronavirus crisis: Strengthening confidence and resilience. OECD. https://www.oecd.org/ctp/tax-policy/tax-and-fiscal-policy-in-response-to-the-coronavirus-crisis-strengthening-confidence-and-resilience.htm.

Önenli, Ö. (2019). *Emission Reductions and Future of Energy Policies In Turkey. Are Renewables An Alternative?* (Ankara: Middle East Technical University).

Özkan, Oktay. (2020). Oynaklçk Sçramas? COVID-19'un Türkiye Hisse Senedi Piyasas? Üzerindeki Etkisi. *Gaziantep University Journal of Social Sciences (COVID-19 Özel Say?)* **19**: 386–397. doi:10.21547/jss.766890.

Pereira da Silva, L.A. (2020). *Green Swan 2 — Climate Change and Covid-19: Reflections on Efficiency Versus Resilience.* (Paris: Bank for International Settlements).

Phan, P.H. and Wood, G. (2020). Doomsday Scenarios (or the Black Swan Excuse for Unpreparedness). *Academy of Management Perspectives*, **34**(4): 425–433.

Rudebusch G.D. (2019). *Climate Change and the Federal Reserve.* (s.l.: FRBSF).

Say, N.P. and Yucel, M. (2006). Energy consumption and CO_2 emissions in Turkey: Empirical analysis and future projection based on an economic growth. *Energy Policy*, **34**: 3870–3876.

Simianer, H. and Reimer, C. (2021). COVID-19: A "black swan" and what animal breeding can learn from it. *Animal Frontiers*, **11**(1): 57–59.

Steininger, K., Bednar-Friedl, B., Formayer, H. and König, M. (2016). Consistent economic cross-sectoral climate change impact scenario analysis: Method and application to Austria. *Climate Services*, **1**: 39–52.

Taleb, N.N. (2001). *Fooled by Randomness: The Hidden Role of Chance in the Markets in Life.* (London: Texere).

Taleb, N.N. (2007). *The Black Swan: The Impact of the Highly Improbable.* (New York: Random House).

Victor, P.A. (2012). Growth, degrowth and climate change: A scenario analysis. *Ecological Economics,* **84**: 206–212.

Villeroy de Galhau, F. (2019). Climate change: Central banks are taking action. *Banque de France-Financial Stability Review,* **23**: 7–13.

Chapter 7

Forecasting the BIST 100 Index with Support Vector Machines

Kamil Demirberk Ünlü[*,†,¶], Nihan Potas[‡,‖],
and Mehmet Ylmaz[§,]**

†*Department of Mathematics, Atılım University,*
Ankara, Turkey

‡*Department of Healthcare Management, Ankara Hacı*
Bayram Veli University, Ankara, Turkey

§*Department of Statistics, Ankara University,*
Ankara, Turkey

¶*demirberk.unlu@atilim.edu.tr*
‖*nihan_potas@hotmail.com*
**mehmetyilmaz@ankara.edu.tr*

Recent literature shows that statistical learning algorithms are powerful for forecasting financial time series. In this study, we model and forecast the Borsa İstanbul 100 Index by employing the machine learning algorithm, support vector machine. The dataset contains the highest price, lowest price, closing price and volume of the index for the period between July 2020 and June 2021. We utilize three different kernels. The empirical findings show that linear kernel gives the best result with coefficient of determination of 0.91 and root mean square error of 0.0062. The second best is polynomial kernel, and it is followed by radial basis kernel. The study shows that statistical learning algorithms can be thought of as an

---

*Corresponding author.

alternative to classical time series methodology in forecasting financial time series

1. Introduction

Forecasting and modeling of financial instruments is of interest to many researchers, financial investors and economists. The stock prices can be considered as the most interesting among the various financial instruments. The stock index can even be seen as an important indicator to the general economy of a country. For these reasons, studies on stock markets have intensified. The modeling and forecasting attempts of stock markets can be divided into the subgroups of time series methodologies and statistical learning.

In time series analysis, autoregressive moving average (ARIMA) type of models are used to forecast the stock prices. In such methodologies, unit root is a key fact. The investigated time series should be stationary to draw meaningful inferences about the data. Multiple unit root tests are available to check the stationarity. If the dataset is not stationary, then the time series should be made stationary by using an appropriate transformation. Also, the error terms of the model satisfy some important assumptions, such as zero mean and homoscedasticity.

Furthermore, the causal relations between stock markets and other possible variables, such as unemployment rate, interest rate, exchange rates and financial indicators, can be investigated by time series methodology. This causal relation can be investigated in the short run by Granger causality and in the long run by cointegration analysis. In order to use these causal inferences again a time series model should be fitted to data. This model also satisfies the assumptions of the methodology. We refer interested readers to Hamilton (2020) and Brockwell and Davis (2009).

Statistical learning is some more flexible than the above-mentioned methodology. It does not have restrictions on the distribution of the time series or on the error term. It can be used either for forecasting or predictions. The challenge in the statistical learning case can be the hyper-parameters. These are the parameters that are not optimized by the algorithm itself. They are determined by the user or by employing other optimization algorithms. More about statistical learning can be found in Alpaydin (2020), James *et al.* (2013) and Kuhn and Johnson (2013).

In the statistical learning case, the two most common sub-groups are machine learning (ML) algorithms and deep learning (DL) algorithms. The ease of use and flexibility of such algorithms increase their popularity. In this study, we employ support vector machine (SVM) algorithm to forecast the daily price of İstanbul Stock Exchange 100 Index (BIST 100). The SVM has the kernel trick which also can be used to model nonlinearly behaved time series. More information can be found under Methodology.

To the best of our knowledge, the leading studies using SVM on stock markets are done by Kim (2003), Huang *et al.* (2005) and Tay and Cao (2001). Kim (2003) used SVM to forecast the direction of Korea composite index. The features of the algorithm are chosen as the financial indicators. The result of the study shows that the model can predict the direction with 50% accuracy. Huang *et al.* (2005) investigated the price movements of 225 stocks which are trading in Tokyo Stock Exchange. The performance of SVM are compared by different methodologies, such as discriminant analysis and neural networks. The findings of the study showed that SVM outperforms the other investigated algorithms. Tay and Cao (2001) forecasted five different future contracts of Chicago Mercantile Exchange. The results revealed the forecasting power of SVM.

Although there are several studies on forecasting stock prices, we try to summarize some recent ML and DL studies on forecasting stock market indices.

Malagrino *et al.* (2018) proposed to use a Bayesian network model to predict the Sao Paulo Exchange closing direction. The results of the study showed that the proposed method has high performance with mean accuracy of 71% and top accuracy of 78%.

Baek and Kim (2018) proposed modularized forecasting framework for stock markets to forecast stock indices of S&P500 and KOSPI 200. The results showed that the proposed model decreased the mean squared error (MSE) and mean absolute error (MAE) by 54.1% and 32.7%, respectively.

Seo *et al.* (2019) employed hybrid artificial neural networks (ANN) to forecast monthly volatilities of S&P500 index. The model was built on GARCH and Google domestic trends. The results revealed that the model with Google trends outperforms the other investigated models.

Xiao *et al.* (2020) used least squares SVM which is hybridized with autoregressive moving average to forecast stock prices. The results indicate that the proposed model has good performance and can be used as a guidance by market regulators and investors.

Karmiani *et al.* (2019) compared SVM, long short-term memory (LSTM) network and recurrent neural network (RNN). The accuracy of these models are used as the performance criterion. The results show that LSTM has the highest accuracy with the lowest variance.

Başoğlu Kabran and Ünlü (2020) modeled the bubbles of S&P500 by a two-step ML algorithm. Multiple ML algorithms are compared and the results showed that the best performance is achieved by SVM.

The rest of this chapter is organized as follows. In the Methodology section, we summarize SVM and introduce three performance metrics to evaluate the proposed algorithms. Empirical results are given in the Analysis section, and the Conclusion section concludes the study with comparison of kernel functions and some future study ideas.

2.   Methodology

SVM is an ML algorithm, which can be used for forecasting and prediction, introduced by Vapnik and Chervonenkis (1974). It uses decisions boundary to classify the data. The fundamental principal of SVM is to find the best hyperplane which divides the data into proper classes.

Let T represent the set of training data which contain input variables $x_i \in \mathfrak{R}^n$ and output variables $y_i \in \mathfrak{R}$, that is $T = \{(x_i, y_i)\}$ for $i = 1, 2, 3, \ldots, N$. In regression, the main idea is to find a flat function $f(x)$ which has ε distance at most.

In the linear case, the function can be represented as

$$f(x) = \langle w, x \rangle + b \tag{1}$$

where $\langle .,. \rangle$ represents the dot product constant b and weight coefficients w. In order to find the best model, the norm value of w, i.e. $\|w\|^2 = \langle w, w \rangle$ is minimized. This turns the problem into an optimization problem as follows:

$$\min \frac{1}{2}\|w\|^2$$

$$\text{s.t}: \begin{array}{l} y_i - \langle w, x_i \rangle - b \le \varepsilon \\ \langle w, x_i \rangle + b - y_i \le \varepsilon \end{array} \tag{2}$$

Here, C is the penalty parameter which represents the relation between the flatness of $f(x)$. After solving the optimization problem, $f(x)$ can be obtained as

$$w = \sum_{i=1}^{n} (\alpha_i - \alpha_i^*) x_i \tag{3}$$

$$f(x) = \sum_{i=1}^{n} (\alpha_i^* - \alpha_i) \langle x_i, x \rangle + b \tag{4}$$

where α_i, α_i^* are Lagrange multipliers and

$$b = y_i - \langle w, x \rangle - \varepsilon \quad \text{for } 0 \le \alpha_i \le C \tag{5}$$

$$b = y_i - \langle w, x \rangle + \varepsilon \quad \text{for } 0 \le \alpha_i^* \le C \tag{6}$$

For the nonlinear case, with the help of kernels, that is $K(x_i, x_j) = \Phi^T(x_i)\Phi(x_j)$, $f(x)$ can be rewritten as

$$f(x) = \sum_{i=1}^{n} (\alpha_i^* - \alpha_i) K(x_i, x) + b \tag{7}$$

Some possible kernels for the SVM are

$$\text{Linear kernel: } K(x_i, x_j^T) = x_i x_j^T \tag{8}$$

$$\text{Polynomial kernel: } K(x_i, x_j^T) = (x_i x_j^T + 1)^d \tag{9}$$

$$\text{Radial basis kernel: } K(x_i, x_j^T) = \exp\left(-\frac{\left\|x_i - x_j^T\right\|^2}{2\sigma^2}\right) \tag{10}$$

More details about SVM and kernels can be found in Vapnik (2013). Here,

$$R^2 = 1 - \frac{SS_{\text{res}}}{SS_{\text{tot}}} \tag{11}$$

$$\text{MAE} = \frac{1}{n} \sum_{i=1}^{n} \left| y_i - \hat{f}_i \right| = \frac{1}{n} \sum_{i=1}^{n} |e_i| \tag{12}$$

$$\text{RMSE} = \sqrt{\frac{1}{n} \sum_{i=1}^{n} \left(y_i - \hat{f}_i \right)^2} = \sqrt{\frac{1}{n} \sum_{i=1}^{n} e_i{}^2} \tag{13}$$

where $SS_{\text{res}} = \sum_{i=1}^{n} \left(y_i - \hat{f}_i \right)^2 = \sum_{i=1}^{n} e_i{}^2$ and $SS_{\text{tot}} = \sum_{i=1}^{n} (y_i - \overline{y})^2$.

3. Analysis

The dataset of the study contains daily opening price (OPN), closing price (CLS), lowest price (LWS), highest price (HGH) and volume (VOL) of BIST 100 index for the period between July 2020 and June 2021. The dataset is obtained from Yahoo Finance (Yahoo, 2021). The descriptive statistics of the data can be found in Table 1.

Figure 1 represents the closing prices of BIST 100 index. The dataset is divided into two parts as training set and test set. The training set contains 80% of the data, while the test set contains 20% of the data.

The algorithm is trained using the training set and tested using the test set. Before we start the analysis, the whole dataset is transformed by using the natural logarithm. We started our analysis with the linear kernel. We employed grid search to find the best hyper-parameters. The best hyper-parameters for the linear kernel are $C = 1$ and $\varepsilon = 0.0005$.

Figure 2 represents the observations and forecasts on the test set. Under these settings, the $RMSE$ of the model is 0.0062 with MAE of 0.0046. Also, it should be noted that R^2 of the linear kernel is calculated as 0.91. In the next step, we utilized the radial basis kernel. For the best case, the hyper-parameters C, ε and γ are calculated as

Table 1. Descriptive statistics of the variables.

Statistics	OPN	CLS	LWS	HGH	VOL
Count	2.150000e+02	2.150000e+02	2.150000e+02	2.150000e+02	2.150000e+02
Mean	1.337620e+09	1.348083e+09	1.323694e+09	1.336157e+09	3.876837e+09
Standard Deviation	1.667545e+08	1.672684e+08	1.661466e+08	1.662094e+08	1.451222e+09
Minimum	1.030300e+09	1.062100e+09	9.854000e+08	1.034400e+09	2.174590e+07
First Quantile	1.155750e+09	1.163750e+09	1.149100e+09	1.152650e+09	2.710134e+09
Median	1.385700e+09	1.399900e+09	1.373700e+09	1.384400e+09	3.910864e+09
Second Quantile	1.481500e+09	1.493750e+09	1.461750e+09	1.475100e+09	4.826008e+09
Maximum	1.578000e+09	1.589500e+09	1.562100e+09	1.570400e+09	9.457914e+09

Fig. 1.   Time series graph of BIST 100 closing prices.

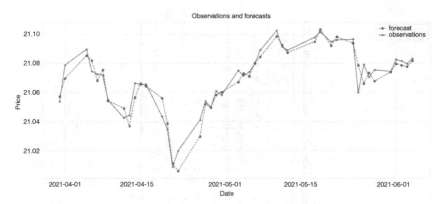

Fig. 2.   Observations and forecasts based on linear kernel.

1000, 0.001 and 5, respectively. Figure 3 represents the observations and forecasts on the test set for the radial basis kernel.

Under these settings, the $RMSE$ of the model is 0.0325 with MAE of 0.0276. This time, a moderate R^2 is calculated as 0.37. The last kernel is the polynomial kernel. The hyper-parameters which give the best forecasting results are $C = 1$, degree $= 2$, $\varepsilon = 0.01$ and $\gamma = 0.1$. Figure 4 represents the observations and forecasts on the test set for the polynomial kernel.

In this case, the $RMSE$ of the model is 0.01203 with MAE of 0.01033. Also, it should be noted that R^2 of the polynomial kernel is calculated as 0.6763.

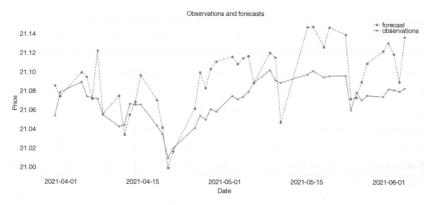

Fig. 3.   Observations and forecasts based on radial basis kernel.

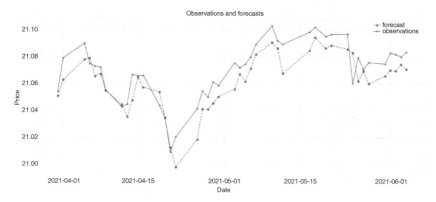

Fig. 4.   Observations and forecasts based on polynomial kernel.

4.   Conclusion

In this study, we aimed to forecast BIST 100 Index by using ML algorithms instead of the time series methodology. We proposed SVM because of the ease of use and flexibility afforded by means of kernel trick. Three different performance measures were used, and they are summarized in Table 2.

According to Table 2, the best model is achieved by the linear kernel. The second best is the polynomial, while the worst performance is achieved by the radial basis kernel. Our analysis shows that, as an alternative, SVM with linear kernel can be employed to model and forecast BIST 100 Index.

Table 2.   Performance measures of the kernels used.

Kernel	MAE	RMSE	R^2
Linear	0.00464	0.0062	0.91
Polynomial	0.0103	0.0120	0.68
Radial	0.0276	0.0325	0.37

In the future, this algorithm can be used to model other stock exchanges. Also, it is possible to consider vector autoregression to model stock index. Further, the investigation of hidden periodicities of the time series can be interesting. For more information about the periodicities, we refer the readers to Akdi and Ünlü (2021) and Okkaoğlu *et al.* (2020). Another interesting topic can be the investigation of the causal relations between stock indices and macroeconomic and microeconomic indicators.

References

Akdi, Y. and Ünlü K.D. (2021). Periodicity in precipitation and temperature for monthly data of Turkey. *Theoretical and Applied Climatology*, **143**(3): 957–968.

Alpaydin, E. (2020). *Introduction to Machine Learning*. (London, England: MIT Press).

Baek, Y. and Kim, H.Y. (2018). ModAugNet: A new forecasting framework for stock market index value with an overfitting prevention LSTM module and a prediction LSTM module. *Expert Systems with Applications*, **113**: 457–480.

Başoğlu, K.F. and Ünlü, K.D. (2020). A two-step machine learning approach to predict S&P 500 bubbles. *Journal of Applied Statistics*, 1–19.

Brockwell, P.J. and Davis, R.A. (2009). *Time Series: Theory and Methods*. (New York, USA: Springer Science & Business Media).

Hamilton, J.D. (2020). *Time Series Analysis*. (Princeton University Press).

Huang, W., Nakamori, Y. and Wang, S.Y. (2005). Forecasting stock market movement direction with support vector machine. *Computers & Operations Research*, **32**(10): 2513–2522.

James, G., Witten, D., Hastie, T. and Tibshirani, R. (2013). *An Introduction to Statistical Learning*, Vol. 112, p. 18 (New York: Springer).

Karmiani, D., Kazi, R., Nambisan, A., Shah, A. and Kamble, V. (2019). Comparison of predictive algorithms: Backpropagation, SVM, LSTM and Kalman filter for stock market. In *2019 Amity International Conference on Artificial Intelligence (AICAI)*. pp. 228–234 (IEEE).

Kim, K.J. (2003). Financial time series forecasting using support vector machines. *Neurocomputing*, **55**(1–2): 307–319.

Kuhn, M. and Johnson, K. (2013). *Applied Predictive Modeling*, Vol. 26 (New York, USA: Springer).

Malagrino, L.S., Roman, N.T. and Monteiro, A.M. (2018). Forecasting stock market index daily direction: A Bayesian network approach. *Expert Systems with Applications*, **105**: 11–22.

Okkaoğlu, Y., Akdi, Y. and Ünlü K.D. (2020). Daily PM10, periodicity and harmonic regression model: The case of London. *Atmospheric Environment*, **238**: 117755.

Seo, M., Lee, S. and Kim, G. (2019). Forecasting the volatility of stock market index using the hybrid models with Google domestic trends. *Fluctuation and Noise Letters*, **18**(01): 1950006.

Tay, F.E. and Cao, L. (2001). Application of support vector machines in financial time series forecasting. *Omega*, **29**(4): 309–317.

Vapnik, V. (2013). *The Nature of Statistical Learning Theory.* (New York: Springer Science & Business Media).

Vapnik, V. and Chervonenkis, A. (1974). *Theory of Pattern Recognition.* (Moscow: Nauka).

Xiao, C., Xia, W. and Jiang, J. (2020). Stock price forecast based on combined model of ARI-MA-LS-SVM. *Neural Computing and Applications*, **32**: 5379–5388.

Yahoo Finance. (2021). BIST 100 (XU100.IS). (Accessed 1 July 2020). https://finance.yahoo.com/quote/XU100.IS?p=XU100.IS&.tsrc=fin-srch.

Chapter 8

Multiple Objective Optimization with Weighted Superposition Attraction–Repulsion (moWSAR) Algorithm

Adil Baykasoğlu

Department of Industrial Engineering,
Dokuz Eylül University, Izmir, Turkey
adil.baykasoglu@deu.edu.tr

Weighted Superposition Attraction–Repulsion (WSAR) algorithm is a recently proposed swarm intelligence-based metaheuristic algorithm that is inspired from superposition principle of physics and attracted movements of agents. WSAR has been applied to many single objective unconstrained and constrained complex optimization problems successfully. In the present study, WSAR is applied to Multiple Objective Optimization (MOO) problems for the first time in the literature. Details of the WSAR algorithm along with applications to MOO problems that are collected from the literature are presented in this study. It is shown that WSAR is competitive and able to generate Pareto optimal solutions.

1. Introduction and Preliminaries

In many real-life settings, decision makers are faced with more than one objective function with conflicts to deal with. Multiple Objective Optimization (MOO) is an important research area in operations

173

research that aims to develop solution methods for this kind of optimization problem. It is possible to define two broad categories of optimization problems based on the number of objective functions, j. If $j = 1$, optimization problems are named as single objective problems; if $j > 1$, optimization problems are defined as MOO problems. We should also note here that problems with more than three objectives (i.e. $j > 3$) are also named as many objective optimization problems in the literature (Adra and Fleming, 2011). MOO problems can be formally represented as follows:

$$\min_{\vec{x} \in R} \vec{f}(\vec{x}) = [f_1(\vec{x}), \dots, f_j(\vec{x})]^T$$
s.t.

$$g_i(\vec{x}) \leq 0 \qquad i = 1, \dots, m$$
$$h_k(\vec{x}) = 0 \qquad k = 1, \dots, n$$
$$L_d \leq x_d \leq U_d \qquad l = 1, \dots, d$$
$$R = \{\vec{x} | g(\vec{x}) \leq 0, \qquad h(\vec{x}) = 0\}$$

where $\vec{f}(\vec{x})$ denotes objective functions vector, j is the number of objective functions to be optimized, $g_i(\vec{x})$ and $h_k(\vec{x})$ denote the set of inequality and equality constraint functions, respectively. m, n and d designate number of inequality and equality constraints and number of decision variables. $[L_d, U_d]$ determines decision variables' upper and lower boundaries. R represents the feasible solution space and $\vec{x} = [x_1, \dots, x_d]^T$ is the Solution Vector (SV). MOO is aimed to find SV^* $(x_1^*, \dots, x_d^*)$ that yields the optimal values for all objective functions that satisfy inequality and equality constraints. However, having more than one objective function changes the notion of optimum in MOO. Because MOO is aimed to find a set of solutions demonstrating tradeoffs among different objective functions rather than a unique optimal solution, as generally a single solution that optimizes all objective functions at once does not exist. The notion of optimum for defining optimal solutions in MOO is the one proposed by Vilfredo Pareto (1896) and the most commonly accepted term to designate optimal solutions is *Pareto optimum*.

Let $SV_a = (SV_{a1}, \dots, SV_{am})$ and $SV_b = (SV_{b1}, \dots, SV_{bm}) \in R$ be two SV, SV_a is supposed to dominate SV_b (denoted as $SV_a \leq SV_b$), if $SV_{ai} \leq SV_{bi}, \forall i = 1, \dots, m$ and $SV_a \neq SV_b$. A point $SV^* \in R$ is called Pareto optimal if there is no $SV \in R$ such that $f(SV)$

dominates $f(SV^*)$. The set of all the Pareto optimal points is named as the *Pareto Set* (*PS*). Set of all Pareto objective vectors, $PF = \{f(SV) \in R | SV \in PS\}$, is named as *Pareto Front* (*PF*). In MOO, the purpose is to find *PS*. Pareto optimality is usually explained with respect to the notion of non-dominated points in the objective space. The goal of MOO is three-fold: (i) The distance of obtained non-dominated solutions to the true Pareto optimal front should be minimized; (ii) A good distribution of the Pareto optimal solutions is wanted; (iii) The spread of the obtained non-dominated solutions should be maximized (Cheng *et al.*, 2012).

Classical optimization techniques are not very effective for solving MOO problems and finding true Pareto optimal solutions for various reasons. For example, the majority of MOO problems are non-convex and consist of disconnected Pareto frontier; finding trade-off solutions may require human-expertise and extensive computations, etc. Moreover, in classical optimization techniques, searching for Pareto optimal solutions simultaneously is generally not the case, instead a scalarization approach (usually weighting sum of objective functions) is utilized in order to turn a MOO problem into a single objective optimization problem. In this approach, finding the right combination of weights and normalization of objective functions with different units is a big challenge. On the other hand, meta-heuristic optimization algorithms are free from the mentioned drawbacks and they are very successful in solving MOO problems. Many metaheuristics algorithms were proposed for solving MOO problems. For example, Baykasoglu *et al.* (1999) proposed the first application of taboo search algorithm for solving MOO problems with success. Fonseca and Fleming (1993) proposed one of the first genetic algorithms for modeling and solving MOO problems. Deb *et al.* (2002) proposed elitist non-dominated sorting genetic algorithm for solving MOO problems with high success. More recently, Dhiman *et al.* (2021) proposed seagull optimization algorithm for solving MOO problems. There are many other metaheuristic approaches and a rich literature on the subject, readers can refer to the following paper for recent approaches (Liu *et al.*, 2020). In this chapter, it is aimed to analyze applicability and efficiency of the Weighted Superposition Attraction–Repulsion (WSAR) algorithm for solving MOO problems for the first time in the literature. WSAR is a recently proposed swarm intelligence-based metaheuristic algorithm that was applied to various complex single objective optimization problems with success (Baykasoglu, 2020). In this chapter, WSAR is modified and adapted for solving MOO problems. Details of the proposed moWSAR are given in the following sections.

2.   The WSAR Algorithm

Baykasoglu (2020) proposed WSAR as a successor of the Weighted Superposition Attraction (WSA) algorithm. WSAR makes use of the superposition principle and attracted movements of particles inspired from classical physics. For that reason, it is classified as a physics-inspired swarm-based metaheuristic algorithm. Differently from its predecessor WSA algorithm, two superpositions are defined in the WSAR algorithm. The first superposition is named as Attractive Superposition (AS), which may attract solution vectors (particles, agents) toward itself. In other words, if a SV has worse fitness value than the AS, then it moves toward the AS. Otherwise, it may move toward the AS with a probability, which compares a random number $(0,1)$ by a value (mp), which is computed as $mp = e^{(f(i)-f(AS))}$, where $f(i)$ is the fitness of the SV i and $f(AS)$ is the fitness of the AS. If the random number is smaller than mp, then the related SV will move toward the AS. Otherwise, it makes random moves by utilizing a random move function. The second superposition is called Repulsive Superposition (RS), which repulses SVs (particles, agents) from its position.

A specially developed biased randomized sampling algorithm that was developed by Baykasoglu (2020) is used to determine superpositions. This algorithm can be used for any type of decision variables without modification except permutation type SVs for preventing repetitions (Baykasoğlu and Senol, 2019). AS and RS in each iteration of WSAR are recomputed as follows [Baykasoglu, 2020]: firstly, an unoccupied vector is created. Afterwards, a random number $(0,1)$ is assigned to each cell of this vector. Subsequently, these random numbers are compared with the formerly calculated rank-based weights (weight columns in Table 1 where τ is set to 0.8) of SVs. Weights are computed as follows: $weight(i) = i^{-\tau}$ where i is the rank of the SV (*SV with the best & the worst fitness gets the 1st rank in determining the AS & RS, respectively*). τ is the main parameter

Table 1. Biased randomized sampling algorithm for computing AS/RS.

Ranking	Solution Vectors (SVs)					Weight $(i) = i^{-\tau}$ $\tau = 0.8$
1	3*	2.21	0*	4.32*	3.11	1.000
2	4	3.22*	1	4.22	4.65	0.574
3	5	4.54	1	3.65	7.65	0.415
4	6	5.65	1	9.67	9.65*	0.329
5	3	7.89	0	1.11	2.42	0.275
6	5	9.11	0	2.45	6.11	0.238
Random numbers	0.841	0.351	0.904	0.564	0.222	
AS	**3**	**3.22**	**0**	**4.32**	**9.65**	

Ranking	Solution Vectors (SVs)					Weight $(i) = i^{-\tau}$ $\tau = 0.8$
1	5	9.11*	0*	2.45*	6.11	1.00
2	3	7.89	0	1.11	2.42	0.574
3	6*	5.65	1	9.67	9.65	0.415
4	5	4.54	1	3.65	7.65	0.329
5	4	3.22	1	4.22	4.65	0.275
6	3	2.21	0	4.32	3.11*	0.238
Random numbers	0.203	0.434	0.876	0.934	0.124	
RS	**6**	**9.11**	**0**	**2.45**	**3.11**	

of the WSAR algorithm that is typically set to 0.8 for achieving a satisfactory balance between intensification and diversification. If an SV's weight is bigger than the formerly set random number, then the corresponding element of the related solution becomes a candidate for the empty position. The roulette wheel technique is used to select one of the candidates as an element of the AS/RS after determining the candidates. This procedure is performed iteratively until complete AS/RS vectors are determined. An illustration for AS/RS vectors' formation is shown in Table 1. Assume that the population comprises six SVs whose 1st dimension is an integer, 3rd dimension is binary and the other dimensions are continuous variables. SVs are ranked from "best-to-worst" and "worst-to-best," $\tau = 0.8$ and random numbers for each dimension are [0.841, 0.351, 0.904, 0.564, 0.222], [0.203, 0.434, 0.876, 0.934, 0.124] for AS/RS vectors, respectively (see Table 1). The values selected from the corresponding dimensions of the SVs undergo roulette wheel selection. These values are emphasized with gray color (see Table 1). For example, considering the 2nd dimension, values selected from the first three SVs enter roulette wheel selection. Assume that the value from the second SV wins the competition. Subsequently, the 2nd dimension of the AS becomes 3.22. Values for the rest of the dimensions are set in a similar manner for AS/RS vectors. For more information, refer to Baykasoglu (2020).

Initialization of SVs: Equation (1) achieves initialization of SVs arbitrarily inside the range of variable boundaries.

$$x_{ij} = x_j^{\min} + \text{rand}(0, 1) * (x_j^{\max} - x_j^{\min}) \tag{1}$$

where $i = 1, 2, \ldots, n\_pop$, $j = 1, 2, \ldots, n.\text{rand}(0, 1)$, x_{ij}, $x_j^{\min}$ and $x_j^{\max}$ designate a random number $\in [0, 1]$, the value of the ith SV at the jth dimension, lower and upper bounds for the jth dimension. n_pop defines number of SVs (population size) and n describes number of decision variables.

Moving of SVs: As pointed out previously, SVs should decide whether to move toward AS or not in each iteration of the WSAR. If AS has better fitness than the compared SV, then SV moves toward it by utilizing Eq. (2), which also helps SV to move away from RS. Else, SV moves randomly by utilizing Eq. (3) (Baykasoglu, 2020).

$$x_{ij}(t+1) = x_{ij}(t) + (\text{rand}(0, 1) * (x_{AS}(t) - |x_{ij}(t)|) - \text{rand}(0, 1)$$
$$*(x_{RS}(t) - |x_{ij}(t)|)) \tag{2}$$

$$x_{ij}(t+1) = x_{ij}(t) + ss(t) * \text{unifrnd}(-1, 1) * |x_{(r1)j}(t) - x_{(r2)j}(t)| \tag{3}$$

where $x_{ij}(t)$ is the value of the position of SV i on dimension j at iteration t; $x_{AS}(t)$ and $x_{RS}(t)$ are the AS and RS at iteration t. $|.\|$ represents absolute function; $x_{(r1)j}(t)$ and $x_{(r2)j}$ represent randomly selected SVs from the population at iteration t. $r1$ and $r2$ are random indexes and $r1 \neq r2 \neq i$; $ss(t)$ is Random Walk Step Sizing (RWSS) function, which starts with an initial step size $ss(0)$ for each variable and updates step sizes as the search evolves by means of a proportional rule (Baykasoglu, 2020). The proportional rule needs a random number and two user-defined parameters, φ_{step}

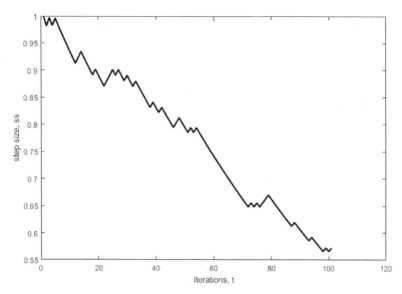

Fig. 1. Behavior of random walk step sizing function ($\varphi_{\text{step}} = 0.7$, $\mu = 0.03$, $ss(0) = 1$).

and μ. Equation (4) is used to compute (t). A graphical illustration that presents behavior of Eq. (4) is given in Fig. 1.

$$ss(t+1) = \begin{cases} ss(t) - e^{-\frac{t}{t+1}} * \varphi_{\text{step}} * ss(t) & \text{if } \text{rand}(0,1) \leq \mu \\ ss(t) + e^{-\frac{t}{t+1}} * \varphi_{\text{step}} * ss(t) & \text{if } \text{rand}(0,1) > \mu \end{cases} \quad (4)$$

The RWSS function has a declining tendency as iterations continue; nevertheless, step size increases in some random instances of search.

Handling constraints: Baykasoğlu and Akpinar (2015) have shown that WSA (predecessor of WSAR) is not sensitive to the selection of constraint handling techniques. Therefore, we usually prefer using parameterless constraint handling approaches within WSAR. The Inverse-Tangent-Constraint-Handling (ITCH) method, which was first proposed by Kim *et al.* (2010) who proved its effectiveness on many optimization problems, is a good selection for WSAR. ITCH assesses SVs by employing Eq. (5).

$$f(\vec{x}) = \begin{cases} \hat{g}(\vec{x}) = g_{\max}(\vec{x}) & \text{if } g_{\max}(\vec{x}) > 0 \\ \hat{f}(\vec{x}) = a \tan[f(\vec{x})] - \pi/2 & \text{otherwise} \end{cases} \quad (5)$$

where $g_{\max}(\vec{x}) = \max_{g_i(\vec{x})}[g_1(\vec{x}), g_2(\vec{x}), g_3(\vec{x}), \ldots, g_m(\vec{x})]$ and $atan[.]$ denotes the inverse tangent. Note that $\hat{g}(\vec{x}) < 0$ for any $\vec{x}$, and thus $\hat{f}(\vec{x}) < \hat{g}(\vec{x})$ is assured.

Handling multiple objectives: The main logic of moWSAR is similar to WSAR with some updates for handling multiple objectives. Dominance ranking and crowding distance evaluation approaches are embedded into WSAR for handling multiple objectives. Differently from WSAR, in moWSAR superiority among the SVs is determined in relation to the non-dominance rank and value of the crowding distance (Δ). SV with the best rank and highest value of Δ is selected as the best SV. In a similar way, the SV with the smallest rank and smallest Δ is selected as the worst SV. During non-domination analyses, first the SVs in the population are controlled and non-dominated SVs are assigned to rank value 1. SVs with rank = 1 are removed from the population and remaining SVs are controlled for non-dominance again and the non-dominated solutions are assigned to rank value 2 and so on. Unless all SVs in the population get a rank, this procedure is repeated. A group of SVs with same rank is named as PF. In moWSAR, between two competing SVs i and j, principally, the SV with a higher rank is preferred. On the other hand, if two SVs have the same rank, then the SV with a higher Δ score is preferred (put in the higher place in WSAR ranking, see Table 1). After determining ranking of all SVs, AS and RS are determined as explained in the previous section (see Table 1). A domination control is also performed between AS and RS. If RS dominates AS, then RS becomes AS and AS becomes RS. In addition, new SVs are generated by using the move equations (Eqs. (2) and (3)) as explained in the previous section with a slight modification. In the moving stage, each SV is compared with AS for domination, if an SV dominates AS, then it moves randomly by making use of Eq. (3), otherwise it moves toward AS by making use of Eq. (2). After moving all SVs, they are combined with the initial population. Consequently, a set of $2 * n\_pop$ solutions is formed. These SVs are ranked once more and the Δ value for every SV is calculated. Based on the new ranking and Δ value n_pop good SVs are selected for the next iteration.

In moWSAR, an external archive of Pareto optimal SVs is kept. After each iteration of moWSAR this archive is updated by eliminating dominated solutions and solutions with smaller Δ scores.

Calculating the crowding distance, Δ: The crowding distance (Δj) is an approximation of the density of *SV*s in the neighborhood of a particular *SV j* (Deb *et al.*, 2002; Rao *et al.*, 2016). For a specific *PF*, let $l = |PF|$ then for each member in *PF*, Δ is computed as follows (Rao *et al.*, 2016):

- Initialize $\Delta j = 0$.
- Sort all *SV*s in the *PF* set in the worst order of objective-function value f_m.
- In the sorted list of mth objective, assign infinite crowding distance to *SV*s at the extremes of the sorted list (i.e. $\Delta 1 = \Delta l = \infty$), for $j = 2$ to $(l - 1)$, compute Δj as follows:

$$\Delta_j = \Delta_j * \frac{f_m^{j+1} - f_m^{j-1}}{f_m^{\max} - f_m^{\min}} \qquad (6)$$

where j shows an *SV* in the sorted list, f_m is the objective-function value of mth objective of jth *SV*, and $f^{\max}$ and $f^{\min}$ are the highest and lowest values of the mth objective function in the current population. Δ is calculated in the similar way for all *SV*s in all *PF*s.

The main steps of the moWSAR algorithm are presented in Table 2.

3. Example Application

The moWSAR algorithm is implemented in MATLAB R2019b. An example test problem is solved on a *PC* with 4 Ghz processor and 32 GB RAM. moWSAR requires few parameters to be tuned. Actually, τ is the only algorithm-specific parameter of moWSAR, which controls the composition of attractive and repulsive superpositions. Reducing τ offers roughly equal chances to candidate *SV*s to be part of attractive and repulsive superpositions. Increasing the value of τ usually leads to a greedy search-like behavior in moWSAR. Based on the previous computational studies and extensive experimental experiences on the predecessors of moWSAR (WSA & WSAR) (Baykasoğlu and Akpinar, 2015; Baykasoğlu, 2020) a value around 0.8 is found suitable for τ. Additionally, random walk step sizing parameters, φ_{step} and μ that are set to 0.003 and 0.7, respectively.

Table 2.   The main steps of the moWSAR algorithm.

- *Initialization*: Set parameters of moWSAR (τ, n_pop, φ_{step}, μ, $Pareto\_Archive\_Size$)
- Generate initial SVs randomly
- Compute objective function values of the initial SVs
- Non-dominated sorting and crowding distance computation for the initial SVs
  $Iteration\_number = 1$
  while $Iteration\_number <= Maximum\_iteration$

  o Rank SVs based on "non-dominated sorting and crowding distance" from "best-to-worst" and "worst-to-best" for determining AS and RS (see Table 1)
  o Assign weights to SVs by considering their ranks (see Table 1)
  o Determine two superpositions, AS & RS, for moving SVs (see Table 1) and calculate their objective function values
  o Evaluate superpositions (AS & RS) based on domination. If RS dominates AS, designate it as AS and vice versa
  o Compare SVs with AS based on domination. If SV dominates AS, move it randomly by using Equation 3, otherwise move it by using Eq. (2)
  o Compute objective function values for each SV
  o Combine new population of SVs with the previous population of SVs
  o Non-dominated sorting and crowding distance computation for the combined population
  o Select n_pop number of good SVs for the next iteration
  o Update Pareto archive by putting non-dominated solutions into it

  $Iteration\_number = Iteration\_number + 1$
  end while

Moreover, a population size (n_pop) around 50 is found adequate. Number of iterations in moWSAR is set to 1500.

Viennet's test problem (Viennet *et al.*, 1996), which has three objective functions and a discrete Pareto-optimal front, is selected for testing the moWSAR. This problem is a complex MOO problem and frequently used by researchers for testing their algorithms. The problem is formally shown by Eqs. (7)–(9).

$$f_1 = 0.5 * (x_1^2 + x_2^2) + \sin(x_1^2 + x_2^2) \tag{7}$$

$$f_2 = \frac{(3 * x_1 - 2 * x_2 + 4)^2}{8} + \frac{(x_1 + x_2 + 1)^2}{27} + 15 \tag{8}$$

$$f_3 = \frac{1}{x_1^2 + x_2^2 + 1} - 1.1 * \exp(-x_1^2 - x_2^2) \tag{9}$$

where $-3 \leq x_1 x_2 \leq 3$. In order to test the quality of the obtained solutions, the inverted generational distance (IGD) is employed as a performance measure. IGD is extensively used as a performance measure in the MOO literature that shows the convergence and diversity of the obtained Pareto optimal solution set. IGD is formally given by Eq. (10) (Li and Wang, 2019).

$$IGD(P^*, P) = \left(\sum_{x^* \in P^*} d(x^*, P) \right) / |P^*| \qquad (10)$$

where P^* is a set of uniformly distributed Pareto optimal points in the Pareto frontier and P is a non-dominated frontier. $d(x^*, P)$ is minimum distance between x^* and any point in P, $|P|$ is cardinality of P^*. True Pareto frontier needs to be known in order to use the IGD measure.

moWSAR is run with the above-defined parameters and 650 Pareto optimal solutions are found as depicted in Fig. 2. In Fig. 2, we also plotted the true Pareto optimal solutions. As it can be seen from this figure, moWSAR is successful in reaching the true Pareto

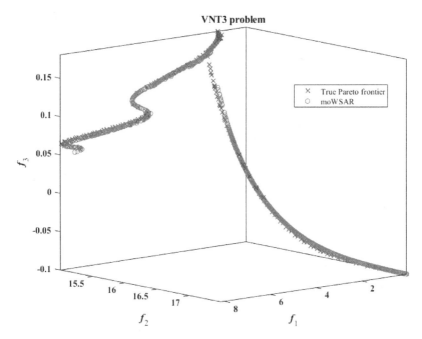

Fig. 2. Pareto frontier generated by utilizing moWSAR for Viennet's problem.

optimal frontier. The average IGD metric after 10 replications is computed as $2.5487E$-1 with a standard deviation of $2.7E$-2. This is an average performance in comparison to reported results. But, we need to note here that moWSAR is still under development stage and extensive computational results will be published in the future.

4. Conclusions

In this work, we presented the moWSAR algorithm that is a multiple objective extension of a recently developed swarm intelligence-based metaheuristic optimization algorithm known as WSAR. moWSAR is tested on a complex three-objective optimization problem. It is observed that it has a promising performance and ability to handle MOO problems. moWSAR is still under development and we aim to apply it to other MOO problems and investigate its performance in detail in future studies.

References

Adra, S.F. and Fleming, P.J. (2011). Diversity management in evolutionary many-objective optimization. *IEEE Transactions on Evolutionary Computation*, **15**(2): 183–195.

Baykasoglu, A. (2020). Optimizing cutting conditions for minimizing cutting time in multi-pass milling via weighted superposition attraction-repulsion (WSAR) algorithm. *International Journal of Production Research*, https://doi.org/10.1080/00207543.2020.1767313.

Baykasoglu, A. and Akpinar, S. (2015). Weighted Superposition Attraction (WSA): A swarm intelligence algorithm for optimization problems — Part 2: Constrained optimization. *Applied Soft Computing*, **37**: 396–415.

Baykasoglu, A. and Senol, M.E. (2019). Weighted superposition attraction algorithm for combinatorial optimization. *Expert Systems with Applications*, **138**: 112792.

Baykasoglu, A., Owen, S., and Gindy, N. (1999). A taboo search based approach to find the Pareto optimal set in multiple objective optimisation. *Engineering Optimization*, **31**(6): 731–748.

Cheng, S., Shi, Y. and Qin, Q. (2012). On the performance metrics of multiobjective optimization. Y. Tan, Y. Shi, and Z. Ji (Eds.): ICSI 2012, Part I, LNCS 7331. pp. 504–512 (Springer-Verlag: Berlin, Heidelberg).

Deb, K., Pratap, A., Agarwal, S. and Meyarivan, T. (2002). A fast elitist non-dominated sorting genetic algorithm: NSGA-II. *IEEE Transactions on Evolutionary Computation*, **6**(2): 182–197.

Dhiman, G., Singh, K.K., Soni, M., Nagar, A., Dehghani, M., Slowik, A., Kaur, A., Sharma, A., Houssein, E.H., Cengiz, K. (2021). MOSOA: A new multi-objective seagull optimization algorithm. *Expert Systems with Applications*, **167**: 114150.

Fonseca, C.M. and Fleming, P.J. (1993). Genetic algorithm for multiobjective optimization: Formulation, discussion and generalization. In *Proceedings of the fifth International Conference on Genetic Algorithms*, pp. 416–423 (San Mateo: Morgan Kauffman Publishers).

Kim, T.H., Maruta, I. and Sugie, T. (2010). A simple and efficient constrained particle swarm optimization and its application to engineering design problems. *Proceedings of the Institution of Mechanical Engineers, Part C: Journal of Mechanical Engineering Science*, **224**(2): 389–400.

Li, H. and Wang L. (2019). A self-organizing map based hybrid chemical reaction optimization algorithm for multiobjective optimization. *Applied Intelligence*, **49**: 2266–2286.

Liu, Q., Li, X., Liu, H. and Guo, Z. (2020). Multi-objective metaheuristics for discrete optimization problems: A review of the state-of-the-art. *Applied Soft Computing*, **93**: 106382.

Pareto, V. (1896). Cours D'Economie Politique. Volume I and II. (F. Rouge: Lausanne).

Rao, R.V., Rai, D.P., Ramkumar, J. and Balic, J. (2016). A new multi-objective Jaya algorithm for optimization of modern machining processes. *Advances in Production Engineering & Management*, **11**(4): 271–286.

Vlennet, R., Fonteix, C. and Marc, I. (1996). Multicriteria optimization using a genetic algorithm for determining a Pareto set. *International Journal of Systems Science*, **27**(2): 255–260.

Chapter 9

Time Series Modeling with Deep Neural Networks

Çağatay Bal[*,†,§] **and Çağdaş Hakan Aladağ**[‡,¶]

[†]*Department of Statistics, Muğla Sıtkı Kocman University, Muğla, Turkey*

[‡]*Department of Statistics, Hacettepe University, Ankara, Turkey*

[§]*cagataybal@mu.edu.tr*

[¶]*aladag@hacettepe.edu.tr*

Deep neural networks are the latest among powerful artificial intelligence tools. As advanced forms of artificial neural networks, deep nets can be used in various fields and also time series forecasting. Time series forecasting is a major domain which extends to almost all problem-wise applications. Because of this reason, powerful tools as deep networks have become the perfect tools with their modular structure for time series forecasting. In this study, starting from shallow neural networks to advanced deep networks, including convolutional nets and long short-term memories, in-depth analytics are investigated and their results are given with applications and Python codes.

---

[*]Corresponding author.

1.   Shallow Neural Networks

Shallow neural networks also known as artificial neural networks (ANNs) are the basic form of deep networks, and the fundamentals of ANNs are based on mimicking the biological neuron structure and communication features (Bal *et al.*, 2016). The primary goal of ANNs is adjusting the connections between neurons, what are called "weights," to understand and process so as to "learn" the behavior of the data with their data-driven characteristic. Training is one of the main processes of ANNs, which is made possible through "learning algorithms," such as backpropagation (BP), soft computing or heuristic algorithms (Aladağ, 2011).

In Fig. 1, the flow of BP is given. As seen in Fig. 1, errors of predictions are obtained and weights are updated as it backpropagates.

Creating and training processes of ANNs consist of many different essential and additional steps resulting in its super-modular structure which is called "architecture." Architecture consists of three main parts called "layers," which are input layer, hidden layer and output layer. These layers contain "neurons" — a process unit of ANNs. Basically, a network only has one input and output layer but can have many hidden layers according to its design. ANNs usually have very complex mathematical forms; therefore, graphical representations of architectures, e.g. shown in Fig. 2, are used more often than topological presentations of networks.

The third important part of ANNs is called "activation functions." These functions are used in order to activate related neurons of each group of data points to obtain predictions. There are two well-known traditional activation functions used in ANNs: "Hyperbolic Tangent Sigmoidal" (tanh) and "Logarithmic Sigmoidal" (logsig). Without activation functions, ANNs cannot decide which neurons to "fire-up" because of information vanishing throughout the calculations. The shapes of these functions can be seen in Fig. 3.

These functions are used in hidden layers of a network, and a "linear" activation function is commonly preferred in the output layer to present predictions as they are.

In Fig. 4, the fundamental structure of ANNs can be seen.

The learning algorithms, architecture and activation functions altogether form the ANNs, an iteration-based computing system which performs well and satisfies expectations in multi-tasking problems. Same as other methods, ANNs also have some advantages

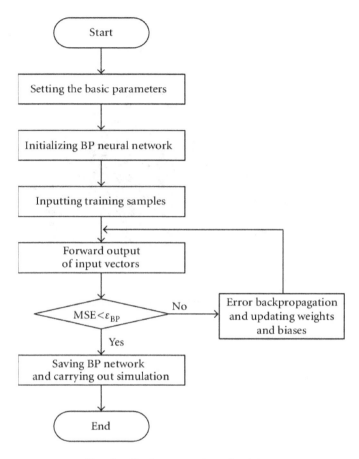

Fig. 1. Backpropagation algorithm.

and disadvantages. Having no assumptions before modeling is one the most important advantage of ANNs. Other advantages include workability with low number of data, goal-driven characteristics and modularity to adopt any shape of modeling task. However, naturally ANNs have disadvantages too, such as random weight initialization at the beginning of a training process, which leads to acquiring different ANNs for each training process. This situation generates an additional need of finding the best ANN among candidates or strategies to acquire the most proper ANN with optimal parameter design. Being a black box is one of the most important characteristics, regarded as a disadvantage of ANNs, which means that there is not a direct explainable relationship between dependent and independent

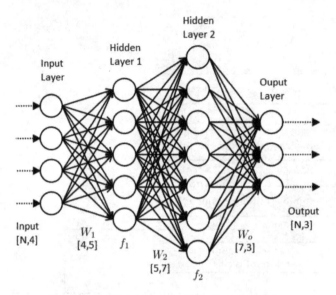

Fig. 2.   Artificial neural network architecture.

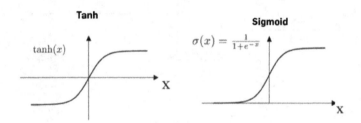

Fig. 3.   Activation functions.

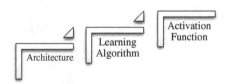

Fig. 4.   Fundamental structure of ANNs.

variables or inputs and outputs. Another disadvantage of ANNs is its data-driven characteristics, which means that trained ANNs are often valid for related data; therefore, generalization is nearly impossible for any ANN applications.

Time series forecasting is a specific field in the modeling literature. For time series forecasting with ANNs, a special form of ANNs called "Feed Forward Neural Networks" (FFNN) are the most useful ones. The interconnected structure among all neurons and one-way signal flow makes this network more suitable for time series forecasting than other types of ANNs.

2. In-Depth Analytics of Shallow Neural Networks

ANNs are considered as universal approximators (Cybenko, 1989) in their very basic form. However, thanks to their modular structures, ANNs can be designed specifically for any kind of problem to increase the performance of the network. To accomplish this, experience in the field and expertise on data are required. As long as a researcher has a good understanding of the data, along with the experience on ANNs, improvements in results are inevitable even with little effort.

The first example is about the XOR problem which was a challenging problem back in the 1950s with "Perception." Perceptron is a primitive version of ANNs consisting of only one neuron and can only be used as a linear separator (Rosenblatt, 1958). Until the 1980s, pessimistic reviews and perspectives led researchers to avoid the topic due to its uselessness (Minsky and Papert, 1969). Their argument was that the Perceptron is not even capable of solving XOR problems in its form.

Figure 5 shows the XOR problem. Naturally, a single linear line is not sufficient to separate two different set of values. A single

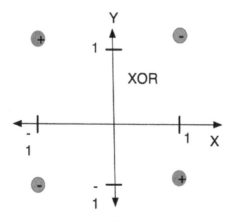

Fig. 5.   XOR problem.

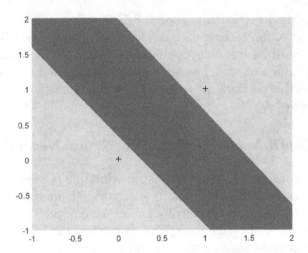

Fig. 6.   Solution to XOR problem.

Perceptron is surely not capable of solve this problem, but if one more neuron can be added to the Perceptron or two Perceptrons are utilized at the same time, the XOR problem can be solved easily.

As shown in Fig. 6, modifying the architecture parameter of Perceptron is enough to solve the XOR problem.

Another example is deciding more accurate parameter settings to shorten the training time, scale-dependent result representation and even improving performance. There are different ways to accomplish this.

Among all BP algorithms, Levenberg–Marquardt algorithm is known for having fast convergence rate because it uses Jacobian matrix to approximate Hessian matrix, which saves considerable time (Levenberg, 1944; Marquardt, 1963). Apart from Levenberg–Marquardt, there are various other algorithms, such as the gradient descent (GD), gradient descent with momentum (GDM), variable learning rate gradient descent with momentum (GDX), Polak–Ribiere conjugate gradient descent (CGP) and quasi-Newton (BFGS). These algorithms may have superiority over one another for different datasets. In Table 1, different results for different datasets are given for comparison (Dao and Vemuri, 2002).

As seen in the table, different algorithms perform differently on each dataset.

Table 1. BP results.

Gradient Descent (GD)

Topology: $\{20, 10, 1\}$	Nbr Epoch $= 1000$
$\alpha = 0.01$	Nbr Flops $= 5$ Gflops
MSE $= 4.5 \exp(-5)$	Error Percentage $= 0$

GD with Momentum (GDM)

Topology: $\{20, 10, 1\}$	Nbr Epoch $= 1000$
$\alpha = 0.01$, $\mu = 0.75$	Nbr Flops $= 5$ Gflops
MSE $= 60$	Error Percentage $= 100$

Variable Learning Rate GD with Momentum (GDX)

Topology: $\{20, 10, 1\}$	Epoch $= 1000$
$\alpha = 0.01$, $\mu = 0.75$, $\beta = 0.7$ and 1.05	Nbr Flops $= 8.4$ Gflops
MSE $= 4.5 \exp(-4)$	Error Percentage $= 0$

Conjugate Gradient Descent (CGP)

Topology: $\{20, 6, 1\}$	Nbr Epoch $= 35$
	Nbr Flops $= 0.23$ Gflops
MSE $= \exp(-5)$	Error Percentage $= 0$

Quasi-Newton (BFGS)

Topology: $\{20, 6, 1\}$	Nbr Epoch $= 20$
	Nbr Flops $= 0.2$ Gflops
MSE $= 1.2 \exp(-6)$	Error Percentage $= 0$

Gradient Descent (GD)

Topology: $\{24, 10, 1\}$	Nbr Epoch $= 1000$
$\alpha = 0.01$	Nbr Flops $= 5.8$ Gflops
MSE $= 1.1 \exp(-4)$	Error Percentage $= 0$

GD with Momentum (GDM)

Topology: $\{24, 10, 1\}$	Nbr Epoch $= 1000$
$\alpha = 0.05$, $\mu = 0.75$	Nbr Flops $= 5.8$ Gflops
MSE $= 2.28 \exp(-4)$	Error Percentage $= 0$

Variable Learning Rate GD with Momentum (GDX)

Topology: $\{24, 10, 1\}$	Epoch $= 1000$
$\alpha = 0.05$, $\mu = 0.75$, $\beta = 0.7$ & 1.05	Nbr Flops $= 1.5$ Gflops
MSE $= 5.1 \exp(-4)$	Error Percentage $= 0$

Conjugate Gradient Descent (CGP)

Topology: $\{24, 6, 1\}$	Nbr Epoch $= 1000$
	Nbr Flops $= 1.5$ Gflops
MSE $= \exp(-5)$	Error Percentage $= 0$

(Continued)

Table 1.   (*Continued*)

Quasi-Newton (BFGS)	
Topology: $\{24, 6, 1\}$	Nbr Epoch $= 45$
	Nbr Flops $= 0.6$ Gflops
MSE $= \exp(-5)$	Error Percentage $= 0$
Gradient Descent (GD)	
Topology: $\{28, 16, 1\}$	Nbr Epoch $= 1000$
$\alpha = 0.01$	Nbr Flops $= 21$ Gflops
MSE $= 2.3 \exp(-5)$	Error Percentage $= 0$
GD with Momentum (GDM)	
Topology: $\{28, 18, 1\}$	Nbr Epoch $= 1000$
$\alpha = 0.05, \mu = 0.75$	Nbr Flops $= 15$ Gflops
MSE $= 2.3 \exp(-4)$	Error Percentage $= 0$
Variable Learning Rate GD with Momentum (GDX)	
Topology: $\{28, 12, 1\}$	Epoch $= 1000$
$\alpha = 0.05, \mu = 0.75, \beta = 0.7$ & 1.05	Nbr Flops $= 7.5$ Gflops
MSE $= \exp(-3)$	Error Percentage $= 0$
Conjugate Gradient Descent (CGP)	
Topology: $\{28, 10, 1\}$	Nbr Epoch $= 45$
	Nbr Flops $= 0.6$ Gflops
MSE $= 1.9 \exp(-3)$	Error Percentage $= 0$
Quasi-Newton (BFGS)	
Topology: $\{28, 8, 1\}$	Nbr Epoch $= 30$
	Nbr Flops $= 0.6$ Gflops
MSE $= \exp(-5)$	Error Percentage $= 0$

There are also newly developed optimizers, such as "Adam" optimizer, which especially work well in deep-learning applications (Kingma and Ba, 2014). The Adam optimizer is an extended version of stochastic gradient descent algorithm which performs well especially on sparse data.

Root mean squared error (RMSE) is a well-known error measure which is widely accepted and used. The formula of RMSE is given as

$$\text{RMSE} = \sqrt{\sum_{i=1}^{n} \frac{(\hat{y}_i - y_i)^2}{n}} \tag{1}$$

However, this simple error measure is scale-dependent which means the data's scale has a direct effect on the RMSE, which makes cross comparison between different datasets impossible. At this point, data normalization can be used to set the data in the [0,1] range to make it suitable for comparison through RMSE manner; also, normalization helps the training process to shorten the time. The normalization function is as follows:

$$z_i = \frac{x_i - \min(x)}{\max(x) - \min(x)} \tag{2}$$

Most of the time, normalization is confused with "standardization." Standardization in statistics refers to the conversion of the data distribution to "Standard Normal Distribution," which is not actually done in normalization. To understand the differences better, the formula of standardization is also given as follows:

$$z = \frac{x - \mu}{\sigma} \tag{3}$$

where z is the standardized observation, μ is the mean value of the data and σ is the standard deviation value of the data.

Objective functions are used in the training process of ANNs, which is obviously an optimization phenomenon. Backpropagation algorithms are what are called learning algorithms based on gradients and therefore derivatives. What if objective function is not differentiable?

$$f(x, y) = \begin{cases} \dfrac{x^2 y}{x^2 + y^2} & \text{if } x^2 + y^2 > 0 \\ 0 & \text{if } x = y = 0 \end{cases} \tag{4}$$

The formula above is a non-differentiable function at $x = y = 0$.

In this case, it is impossible to use backpropagation algorithms to train a network. Instead of using backpropagation algorithms, heuristic algorithms, such as particle swarm optimization (PSO), tabu search and bee colony, can be utilized (Zhang *et al.*, 2007).

Data may contain outliers in problematic scale. This situation becomes very crucial in the learning process. The reason is that the aggregation function in each neuron of network forms around a "mean" operator.

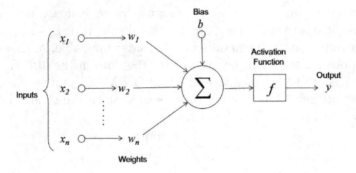

Fig. 7.   Neuron aggregation function.

As shown in Fig. 7, an aggregation function inside of a neuron can be seen clearly.

As it is known from statistics, the mean operator is highly sensitive to outliers which will become so effective on training of ANN therefore poor performance of network. To avoid this situation more in-depth modification of changing the mean operator with median in aggregation function is required (Aladağ *et al.*, 2014). With this solution, the network will not be as vulnerable to outliers, so it attains a more "robust" form. The following formulas show the default and median forms of neuron aggregation functions:

$$\text{net}(x_j, w_j) = \sum_{j=1}^{N} w_j x_j + w_0$$

$$\text{net} = \text{Median}(w_1 x_1, w_2 x_2, \ldots, w_N x_N, w_0) \qquad (5)$$

Overfitting is a major issue in ANNs. Sometimes, a network avoids "learning" the data and instead mimics the observations one to one. When this happens, the network shows poor performance on "out-of-sample" predictions. To avoid overfitting, there are different strategies that can be utilized, such as dropout, regularization, dataset partition of training, testing and validation and training many networks in a multiple parameter design space to find the best network in a trial-and-error manner (Bal and Demir, 2017).

Trial and error is a very common technique among the aforementioned. Unlike parametric methods, which have a unique solution, nonparametric methods could have infinite possible solutions for a problem. Having numerous solutions requires strategies such as selecting the best among all solutions. The simple but effective solution is widely used and also specialized functions has been proposed (Bal, 2021).

Dropout is a basic but effective technique to reduce overfitting in ANNs. It basically "drops out" some neurons randomly to create a challenge for ANNs during the training period. This punishment makes ANNs work harder to train and therefore reduce overfitting. In Fig. 8, a dropout configuration can be seen for better understanding.

Regularization is a technique which generally refers to "penalizing" the analysis using functions. Penalization is basically based on functional conditions to prevent unwanted situations. One of the very first examples in the literature is Akaike information criterion (AIC) (Akaike, 1974). This criterion has a second term in its function to penalize parameter number in a model. Its design tries to penalize models with large number of parameters which is not preferred mostly in ARIMA methods (Box and Jenkins, 1976). The AIC function can be seen as follows:

$$AIC = 2k - 2\ln\left(\hat{L}\right) \tag{6}$$

where k is the number of estimated parameters in the model and $\hat{L}$ is the maximum value of the likelihood function for the model.

Vanishing gradients is another major issue in ANNs. This phenomenon occurs as a multiplication of gradients and weights that go

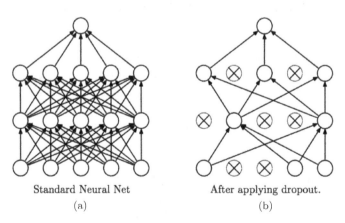

Standard Neural Net
(a)

After applying dropout.
(b)

Fig. 8.   Dropout network.

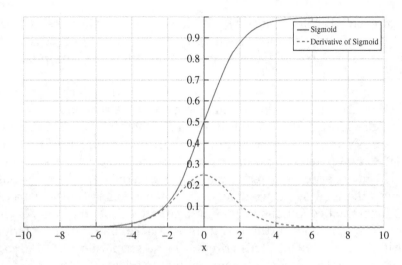

Fig. 9.   Sigmoidal activation function derivatives.

deep into a network. After a long chain of multiplications, the gradients become extremely low, which becomes meaningless for learning because all the information disappears. For better understanding, a visual representation of vanishing gradients as a relationship between sigmoidal activation functions and its derivatives is given in Fig. 9.

More detailed information about vanishing gradients will be discussed in the section on long short-term memory (LSTM).

3.   Convolutional Neural Networks for Modeling (1-D CNN)

Convolutional neural networks (CNNs) are specialized deep learning methods for image recognition applications (Fukushima, 1980; Goodfellow *et al.*, 2016; Li *et al.*, 2021). This powerful network is able to learn and recognize thousands of images. Though the first studies started way back in the 1990s, the first successful CNN, named AlexNet, was introduced in 2012 (Krizhevsky *et al.*, 2012). AlexNet is a pioneer of using "ReLu" activation function in a network to mitigate the vanishing gradients problem, while also using dropout method to avoid overfitting, thus making it suitable for use in graphical processing units (GPUs) to train the network.

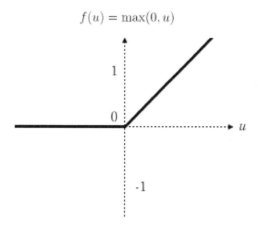

$$f(u) = \max(0, u)$$

Fig. 10.   ReLu activation function.

The ReLu activation function can be seen in Fig. 10. This function is a revolutionary step for deep learning. The reason is that, apart from other tanh and logsig functions, ReLu makes calculations very easy and less time-consuming. When considering deep networks having numerous layers and neurons in their architecture, traditional activation functions practically make training process enormously long. The other advantage of ReLu is disabling irrelevant neurons functionally when the related data flow in the network.

Moreover, although CNNs are designed to work with two- or three-dimensional images or objects, one-dimensional (1-D) networks can be created to utilize time series forecasting. A 1-D CNN can combine 1-D data with the powerful CNN properties. A 1-D CNN developed by Kiranyaz *et al.* (2021) studies especially with datasets such as those from biomedical, early diagnosis, structural health monitoring, anomaly detection and identification in power electronics and electrical motor fault detection. Working with 1-D CNNs is relatively easy in terms of complexity of network, computational load and time requirement. The superiority of 1-D CNNs over shallow networks is undeniable in most cases of time series forecasting analysis. The reason is that CNN is basically designed to work with "spatial" data and time series are also spatial which makes them absolutely in accordance.

An example of time series forecasting with Python code and results of 1-D CNN are as follows (Fig. 11).

```
# Part 1 - Data Preprocessing
import numpy as np
import matplotlib.pyplot as plt
import pandas as pd


dataset_train = pd.read_csv('AirPassengers_train.csv')
training_set = dataset_train.iloc[:, 1:2].values


from sklearn.preprocessing import MinMaxScaler
sc = MinMaxScaler(feature_range = (0, 1))
training_set_scaled = sc.fit_transform(training_set)


X_train = []
y_train = []
for i in range(14, 130):
    X_train.append(training_set_scaled[i-14:i, 0])
    y_train.append(training_set_scaled[i, 0])
X_train, y_train = np.array(X_train), np.array(y_train)
# Verinin tekrar şekillendirilip analize hazır hale getirilmesi
X_train = np.reshape(X_train, (X_train.shape[0], X_train.shape[1], 1))


# Part 2 - 1D-CNN
from numpy import array
from keras.models import Sequential
from keras.layers import Dense
from keras.layers import Flatten
from keras.layers.convolutional import Conv1D
from keras.layers.convolutional import MaxPooling1D
```

```python
# 1D-CNN bileşenlerinin oluşturulması

model = Sequential()

# First 1D-CNN layer

model.add(Conv1D(filters=64, kernel_size=2, activation='relu'))

# 1D-CNN layer`s Pooling

model.add(MaxPooling1D(pool_size=2))

# Second 1D-CNN layer

model.add(Conv1D(filters=32, kernel_size=2, activation='relu'))

model.add(MaxPooling1D(pool_size=2))

# Flattening layer


model.add(Flatten())

# Output layer

model.add(Dense(50, activation='relu'))

model.add(Dense(1))

model.compile(optimizer='adam', loss='mse')

model.fit(X_train, y_train, epochs=1000, verbose=0)


# Part 3 - Predictions

dataset_test = pd.read_csv('AirPassengers_test.csv')

real_AirPassengers = dataset_test.iloc[:, 1:2].values


dataset_total = pd.concat((dataset_train['#Passengers'], dataset_test
['#Passengers']),axis =0)

inputs = dataset_total[len(dataset_total) - len(dataset_test) - 14:].values

inputs = inputs.reshape(-1,1)

inputs = sc.transform(inputs)

X_test = []

for i in range(14, 28):

    X_test.append(inputs[i-14:i, 0])

X_test = np.array(X_test)

X_test = np.reshape(X_test, (X_test.shape[0], X_test.shape[1], 1))
```

```
predicted_AirPassengers = model.predict(X_test)
predicted_AirPassengers = sc.inverse_transform(predicted_AirPassengers)


plt.plot(real_AirPassengers, color = 'red', label = 'Gerçek AirPassengers ')
plt.plot(predicted_AirPassengers, color = 'blue',
label ='Tahmin AirPassengers ')
plt.title('AirPassengers Tahminleri')
plt.xlabel('Zaman')
plt.ylabel('AirPassengers')
plt.legend()
plt.show()


from sklearn.metrics import mean_squared_error
from math import sqrt


rmse = sqrt(mean_squared_error(real_AirPassengers, predicted_AirPassengers))
Out[]: 33.60583523107709
```

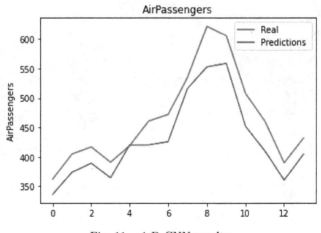

Fig. 11.   1-D CNN results.

4. Long Short-Term Memory (LSTM) Network for Modeling

LSTM networks are the best deep learning tools for long-sequence data, such as time series. Its ability to learn long-time relationship between observations make LSTM a perfect tool for time series forecasting (Hochreiter and Schmidhuber, 1997). Before explaining LSTMs, it is necessary to recognize what recurrent neural networks (RNNs) are first. RNNs are shallow neural networks that have a major distinction from FFNNs architecture wise. RNNs have inner repeatable relationships between neurons in a hidden layer.

As shown in Fig. 12, unlike FFNNs, the hidden layer neurons of RNNs can update itself as per a certain condition. This feature allows RNNs to carry on and learn long-time relationships in time series data. However, this multiplying process has a major downside which is called "Vanishing Gradients" as mentioned before. Gradient vanishing occurs when multiplication of gradients continues many times such that they become meaninglessly low in the learning process.

The first solution to avoid this problem is using different activation functions, such as ReLu. ReLu derivates are either zero or one which makes the learning possible to continue flowing through the network without vanishing.

Also, to overcome this weakness, LSTM uses a more complex neuron type. In LSTM, each neuron has three gates: input, output and forget gates. These gates decide which lagged data will be passed

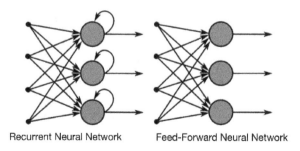

Recurrent Neural Network   Feed-Forward Neural Network

Fig. 12.   RNN architecture.

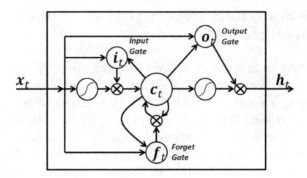

Fig. 13.   LSTM cell.

through to the next cell and which will not be. A typical LSTM neuron can be seen in Fig. 13.

With this in-neuron strategy, LSTM overcomes the vanishing gradients problem and learns long-time relationships in data. Some specially designed LSTMs can even write a movie script (Buder, 2016).

An example of time series forecasting with Python code and results for LSTM are given as follows (Fig. 14).

```python
# Part 1 - Data Preprocessing
import numpy as np
import matplotlib.pyplot as plt
import pandas as pd


dataset_train = pd.read_csv('AirPassengers_train.csv')
training_set = dataset_train.iloc[:, 1:2].values


from sklearn.preprocessing import MinMaxScaler
sc = MinMaxScaler(feature_range = (0, 1))
training_set_scaled = sc.fit_transform(training_set)


X_train = []
y_train = []
for i in range(14, 130):
```

```python
    X_train.append(training_set_scaled[i-14:i, 0])
    y_train.append(training_set_scaled[i, 0])
X_train, y_train = np.array(X_train), np.array(y_train)
# Verinin tekrar şekillendirilip analize hazır hale getirilmesi
X_train = np.reshape(X_train, (X_train.shape[0], X_train.shape[1], 1))


# Part 2 - LSTM
from keras.models import Sequential
from keras.layers import Dense
from keras.layers import LSTM
from keras.layers import Dropout
regressor = Sequential()
# First LSTM layer with `Dropout` regularization
regressor.add(LSTM(units = 50, return_sequences = True,
input_shape = (X_train.shape[1], 1)))
regressor.add(Dropout(0.2))
# Second LSTM layer with `Dropout` regularization
regressor.add(LSTM(units = 50, return_sequences = True))
regressor.add(Dropout(0.2))
# Third LSTM layer with `Dropout` regularization
regressor.add(LSTM(units = 50, return_sequences = True))
regressor.add(Dropout(0.2))
# Fourth LSTM layer with `Dropout` regularization
regressor.add(LSTM(units = 50))
regressor.add(Dropout(0.2))
# Output layer
regressor.add(Dense(units = 1))
regressor.compile(optimizer = 'adam', loss = 'mean_squared_error')


regressor.fit(X_train, y_train, epochs = 500, batch_size = 50)
```

```
# Part 3 - Predictions

dataset_test = pd.read_csv('AirPassengers_test.csv')

real_AirPassengers = dataset_test.iloc[:, 1:2].values


dataset_total = pd.concat((dataset_train['#Passengers'],
dataset_test['#Passengers']),axis =0)

inputs = dataset_total[len(dataset_total) - len(dataset_test) - 14:].values

inputs = inputs.reshape(-1,1)

inputs = sc.transform(inputs)

X_test = []

for i in range(14, 28):

    X_test.append(inputs[i-14:i, 0])

X_test = np.array(X_test)

X_test = np.reshape(X_test, (X_test.shape[0], X_test.shape[1], 1))

predicted_AirPassengers = model.predict(X_test)

predicted_AirPassengers = sc.inverse_transform(predicted_AirPassengers)


plt.plot(real_AirPassengers, color = 'red', label = 'Gerçek AirPassengers ')

plt.plot(predicted_AirPassengers, color = 'blue',

label ='Tahmin AirPassengers ')

plt.title('AirPassengers Tahminleri')

plt.xlabel('Zaman')

plt.ylabel('AirPassengers')

plt.legend()

plt.show()


from sklearn.metrics import mean_squared_error
from math import sqrt


rmse = sqrt(mean_squared_error(real_AirPassengers, predicted_AirPassengers))
Out[]: 36.56466157604625
```

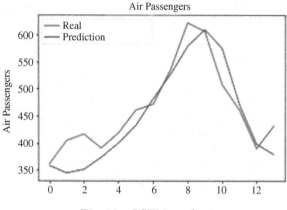

Fig. 14.   LSTM results.

The results show that 1-D CNNs are very successful for time series prediction. It is obvious that convolutional networks, which can be used in many different fields such as not only visual identification but also time series prediction, are an approach that has high potential and is always open to development.

References

Akaike, H. (1974). A new look at the statistical model identification. *IEEE Transactions on Automatic Control*, **19**(6), 716–723. https://doi.org/10.1109/TAC.1974.1100705.

Aladağ, C.H. (2011). A new architecture selection method based on tabu search for artificial neural networks. *Expert Systems with Applications*, **38**(4), 3287–3293. https://doi.org/10.1016/j.eswa.2010.08.114.

Aladağ, C.H., Egrioglu, E. and Yolcu, U. (2014). Robust multilayer neural network based on median neuron model. *Neural Computing and Applications*, **24**(3–4), 945–956. https://doi.org/10.1007/s00521-012-1315-5.

Bal, C. (2021). cbnet (https://www.mathworks.com/matlabcentral/file exchange/67628-cbnet), *MATLAB Central File Exchange*. Retrieved July 18, 2021.

Bal, C. and Demir, S. (2017). Forecasting TRY/USD exchange rate with various artificial Neural Network Models. *TEM Journal*, **6**(1), 11–16. https://doi.org/10.18421/TEM61-02.

Bal, C., Demir, S. and Aladağ, C.H. (2016). A comparison of different model selection criteria for forecasting EURO/USD exchange rates by feed forward neural network. *International Journal of Computing, Communications & Instrumentation Engineering*, **3**(2), 1–5.

Box, G.E.P. and Jenkins, G.M. (1976). *Time Series Analysis: Forecasting and Control (Revised Edition)*, (San Franciso, CA: Holden-Day).

Buder, E. (2016). *An Algorithm Wrote This Movie, and It's Somehow Amazing*, (No Film School) Retrieved June 13, 2016.

Cybenko, G. (1989). Approximation by superpositions of a sigmoidal function. *Mathematics of Control, Signals and Systems*, **2**(4), 303–314. https://doi.org/10.1007/BF02551274.

Dao, V.N.P. and Vemuri, R. (2002). A performance comparison of different back propagation neural networks methods in computer network intrusion detection. *Differential Equations and Dynamical Systems*, **10**(1&2), 1–7.

Fukushima, K. (1980). Neocognitron: A self-organizing neural network model for a mechanism of pattern recognition unaffected by shift in position. *Biological Cybernetics*, **36**(4), 193–202. https://doi.org/10.1007/BF00344251.

Goodfellow, I., Bengio, Y. and Courville, A. (2016). *Deep Learning*, (MIT Press).

Hochreiter, S. and Schmidhuber, J. (1997). Long short-term memory. *Neural Computation*, **9**(8), 1735–1780.

Kingma, D.P. and Ba, J. (2014). Adam: A Method for Stochastic Optimization. *3rd International Conference on Learning Representations*.

Kiranyaz, S., Avci, O., Abdeljaber, O., Ince, T., Gabbouj, M. and Inman, D. J. (2021). 1D convolutional neural networks and applications: A survey. *Mechanical Systems and Signal Processing*, **151**, 107398. https://doi.org/10.1016/j.ymssp.2020.107398.

Krizhevsky, A., Sutskever, I. and Hinton, G.E. (2012). ImageNet classification with deep convolutional neural networks. In F. Pereira, C.J.C. Burges, L. Bottou, and K.Q. Weinberger (Eds.): *Advances in Neural Information Processing Systems* (Vol. 25), (Curran Associates, Inc.) https://proceedings.neurips.cc/paper/2012/file/c399862d3b9d6b76c8436e924a68c45b-Paper.pdf.

Levenberg, K. (1944). A method for the solution of certain non-linear problems in least squares. *Quarterly of Applied Mathematics*, **2**(2), 164–168. http://www.jstor.org/stable/43633451.

Li, Z., Liu, F., Yang, W., Peng, S. and Zhou, J. (2021). A survey of convolutional neural networks: Analysis, applications, and prospects. *IEEE Transactions on Neural Networks and Learning Systems*, 1–21. https://doi.org/10.1109/tnnls.2021.3084827.

Marquardt, D.W. (1963). An Algorithm for Least-Squares Estimation of Nonlinear Parameters. *Journal of the Society for Industrial and Applied Mathematics*, **11**(2), 431–441. https://doi.org/10.1137/0111030.

Minsky, M. and Papert, S. (1969). *Perceptrons: An Introduction to Computational Geometry*, (MIT Press).

Rosenblatt, F. (1958). The perceptron: A probabilistic model for information storage and organization in the brain. *Psychological Review*, **65**(6), 386–408.

Zhang, J.R., Zhang, J., Lok, T.M. and Lyu, M.R. (2007). A hybrid particle swarm optimization-back-propagation algorithm for feedforward neural network training. *Applied Mathematics and Computation*, **185**(2), 1026–1037. https://doi.org/10.1016/j.amc.2006.07.025.

Chapter 10

An Extension of the Inverse Gaussian Distribution

Talha Arslan

*Department of Econometrics, Van Yüzüncü Yıl University,
Van, Turkey*

mstalhaarslan@yyu.edu.tr

In this study, an α-monotone extension of the inverse Gaussian (αIG) distribution is introduced. Then, the method of moments estimations for the parameters of the αIG distribution is provided. A real dataset is used to show the fitting performance of the αIG distribution. The results show that the αIG distribution fits the corresponding dataset better than the IG distribution if the well-known goodness-of-fit statistics are taken into account. Note that the αIG distribution is defined as a general class of the IG distribution by adding a new shape parameter. It can be considered an alternative to the IG distribution in modeling data from different areas of science.

1. Introduction

The inverse Gaussian (IG) distribution has been used for modeling data since the paper by Tweedie (1945). The IG distribution arises from the density of the first passage time of the Brownian motion;

see Tweedie (1957a;b) and Folks and Chhikara (1978). The IG distribution has the probability density function (PDF)

$$f_X(x; \lambda, \mu) = \left(\frac{\lambda}{2\pi x^3}\right)^{1/2} \exp\left\{-\frac{\lambda(x-\mu)^2}{2\mu^2 x}\right\};$$

$$x > 0, \quad \lambda > 0, \quad \mu > 0$$

and cumulative distribution function (CDF)

$$F_X(x; \lambda, \mu) = \frac{1}{\sqrt{\pi}} \exp\left\{\frac{\lambda}{\mu}\right\} \Gamma\left[\frac{1}{2}, \frac{\lambda}{2} x^{-1}; \frac{\lambda^2}{4\mu^2}\right]$$

$$= \frac{1}{2}\left[\operatorname{Erfc}\left(\sqrt{\frac{\lambda}{2} x^{-1}} - \frac{\frac{\lambda}{2\mu}}{\sqrt{\frac{\lambda}{2} x^{-1}}}\right)\right.$$

$$\left. + \exp\left\{\frac{2\lambda}{\mu}\right\} \operatorname{Erfc}\left(\sqrt{\frac{\lambda}{2} x^{-1}} + \frac{\frac{\lambda}{2\mu}}{\sqrt{\frac{\lambda}{2} x^{-1}}}\right)\right]$$

Here, $\Gamma\left[\frac{1}{2}, \frac{\lambda}{2} x^{-1}; \frac{\lambda^2}{4\mu^2}\right]$ represents the generalized incomplete gamma function defined by Chaudhry and Zubair (1994) and Erfc($\cdot$) denotes the complementary error function. Note that the CDF of the IG distribution can also be expressed by means of the CDF of the standard normal distribution.

The rth raw moment of the IG distribution is

$$\mathbb{E}[X^r] = \exp\left\{\frac{\lambda}{\mu}\right\} \sqrt{\frac{2\lambda}{\pi\mu}} \mu^r K_{\frac{1}{2}-r}\left(\frac{\lambda}{\mu}\right) \tag{1}$$

where $K_s(\cdot)$ denotes the modified Bessel function of the second kind. Then, the first four raw moments of the IG distribution are formulated as

$$\mathbb{E}[X] = \mu$$

$$\mathbb{E}[X^2] = \frac{\mu^2(\lambda + \mu)}{\lambda}$$

$$\mathbb{E}[X^3] = \frac{\mu^3(\lambda^2 + 3\lambda\mu + 3\mu^2)}{\lambda^2}$$

and

$$\mathbb{E}[X^4] = \frac{\mu^4}{\lambda}(\lambda^4 + 6\mu\lambda^3 + 15\lambda^4\mu^2 + 15\lambda\mu^3)$$

In the literature, there exist papers including extensions/ generalizations of the IG distribution. For example, Iwase and Hirano (1990) obtained the power IG distribution, Butler (1998) introduced the generalized IG (GIG) distribution, Mathai and Provost (2011) defined the q-extended IG distribution, Lemonte and Cordeiro (2011) proposed the exponentiated GIG distribution, Vasconcelos *et al.* (2019) obtained the odd log-logistic GIG distribution and Junfeng *et al.* (2020) introduced the extended IG distribution.

There exist many different methods used for generalizing/ extending a distribution. Lee *et al.* (2013) stated that most of these methods are based on adding a new parameter to the baseline distribution. Recently, Jones (2020) considered the distribution of the form

$$T = X \times Y^{\frac{1}{\alpha}} \tag{2}$$

and called it α-monotone density. Here, X and Y are independent random variables following distributions on $\mathbb{R}^+$ and Uniform(0,1), respectively.

The α-monotone density has a simple concept; therefore, this can be used by researchers for introducing modified/generalized/ extended α-monotone distributions; see for example Arslan (2021a).

The α-monotone concept has been considered in a limited number of studies. The motivation of this study comes from this fact; therefore, an α-monotone extension of the IG (αIG) distribution is introduced to contribute to the related literature. The αIG distribution is obtained as a product of IG and $(1/\alpha)$-power of a Uniform(0,1) independent random variable. An earlier version of this study was presented at the *International Conference on Computational Mathematics and Engineering Sciences*; see Arslan (2021b).

2. αIG Distribution

The PDF and CDF of the random variable T having αIG distribution are

$$
\begin{aligned}
f_T(t; \alpha, \lambda, \mu) \\
&= \frac{\alpha}{\sqrt{2\pi}} \sqrt{\lambda} \exp\left\{\frac{\lambda}{\mu}\right\} t^{\alpha-1} \int_t^\infty u^{-\alpha-3/2} \exp\left\{-\frac{\lambda}{2\mu^2}u - \frac{\lambda}{2}u^{-1}\right\} \\
&= \frac{\alpha}{\sqrt{\pi}} \frac{2^\alpha \mu^{2\alpha+1}}{\lambda^\alpha} \exp\left\{\frac{\lambda}{\mu}\right\} t^{\alpha-1} \Gamma\left[-\alpha - 0.5, \frac{\lambda}{2\mu^2}; \frac{\lambda^2}{4\mu^2}\right] \quad (3)
\end{aligned}
$$

and

$$
\begin{aligned}
F_T(t; \alpha, \lambda, \mu) &= F_X(t; \lambda, \mu) + \frac{t}{\alpha} f_T(t; \alpha, \lambda, \mu) \\
&= \frac{1}{\sqrt{\pi}} \exp\left\{\frac{\lambda}{\mu}\right\} \Gamma\left[\frac{1}{2}, \frac{\lambda}{2}x^{-1}; \frac{\lambda^2}{4\mu^2}\right] \\
&\quad + \frac{1}{\sqrt{\pi}} \frac{2^\alpha \mu^{2\alpha+1}}{\lambda^\alpha} \exp\left\{\frac{\lambda}{\mu}\right\} t^\alpha \Gamma\left[-\alpha - 0.5, \frac{\lambda}{2\mu^2}; \frac{\lambda^2}{4\mu^2}\right]
\end{aligned}
$$

$$(4)$$

respectively. Here, $t > 0$, $\alpha > 0$, $\lambda > 0$ and $\mu > 0$. The proofs are not given here for the sake of brevity; see Jones (2020). In Fig. 1, the PDF of the αIG distribution is plotted for certain values of the distribution parameters.

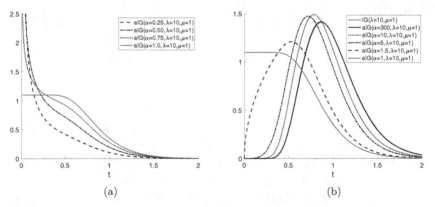

Fig. 1. (a) $\alpha \leq 1$ and (b) $\alpha \geq 1$ density plots (PDFs) of the αIG distribution for different parameter settings.

Note that from the definition given in Eq. (2):

i. $\lim\limits_{\alpha \to \infty} f_T(x; \alpha, \lambda, \mu) = f_X(x; \lambda, \mu)$; see also Fig. 1.

ii. the rth moment of the αIG distribution is formulated as

$$\mathbb{E}[T^r] = \exp\left\{\frac{\lambda}{\mu}\right\} \sqrt{\frac{2\lambda}{\pi\mu}} \mu^r K_{\frac{1}{2}-r}\left(\frac{\lambda}{\mu}\right) \frac{\alpha}{\alpha+r}$$

Here, the result given in (ii) is obtained by the product of the rth moment of the IG distribution given in Eq. (1) and (r/α)th moment of the $U(0,1)$ distribution since random variable X and Y are independent.

3. Parameter Estimation

In this section, the method of moment (MoM) estimations of the parameters of the αIG distribution are provided. The MoM estimators of the parameters α, λ and μ can be obtained by equating the first three theoretical moments to the corresponding sample moments as follows:

$$\mathbb{E}[T] = \left(\frac{\alpha}{\alpha+1}\right)\mu = T_1 \tag{5}$$

$$\mathbb{E}[T^2] = \left(\frac{\alpha}{\alpha+2}\right)\frac{\mu^2(\lambda+\mu)}{\lambda} = T_2 \tag{6}$$

and

$$\mathbb{E}[T^3] = \left(\frac{\alpha}{\alpha+3}\right)\frac{\mu^3(\lambda^2 + 3\lambda\mu + 3\mu^2)}{\lambda^2} = T_3 \tag{7}$$

Here, $T_1 = (1/n)\sum_{i=1}^{n} t_i$, $T_2 = (1/n)\sum_{i=1}^{n} t_i^2$ and $T_3 = (1/n)\sum_{i=1}^{n} t_i^3$. From Eq. (5), the MoM estimator of the parameter μ, i.e. $\hat{\mu}_{\text{MoM}}$, is

$$\hat{\mu}_{\text{MoM}} = \frac{\alpha+1}{\alpha}T_1 \tag{8}$$

After incorporating the $\hat{\mu}_{\text{MoM}}$ into Eqs. (6) and (7), then equating them to zero, the system of equations

$$\frac{(\alpha+1)^2}{\lambda\alpha^2(\alpha+2)}\left(\alpha T_1^3 + \alpha\lambda T_1^2 + T_1^3\right) - T_2 = 0$$

$$\frac{(\alpha+1)^3}{\lambda\alpha^5(\alpha+3)}\left(\alpha^2\lambda^2 T_1^3 + 3\alpha^2\lambda T_1^4 + 3\alpha\lambda T_1^4 + 3\alpha^2 T_1^5 + 6\alpha T_1^5 + 3T_1^5\right) - T_3 = 0$$

$$(9)$$

is obtained. The MoM estimates of the α and λ, i.e. values of the $\hat{\alpha}_{\text{MoM}}$ and $\hat{\lambda}_{\text{MoM}}$, are obtained by solving the system of equations in Eq. (9).

4. Application

In this section, a dataset including 116 observations of daily ozone level measurements in New York for the period May–September 1973 is modeled using the αIG and IG distributions. Note that Nadarajah (2008), who provided this dataset, used truncated inverted beta distribution for modeling purpose; see Table 1 for the corresponding dataset.

The well-known goodness-of-fit statistics Kolmogorov–Smirnov (KS), Anderson–Darling (AD), root mean squared error (RMSE) and coefficient of determination (R^2) are used to compare the fitting performances of the αIG and IG distributions; see Arslan (2021a) for their formulas.

Table 1.   Air pollution data ($n = 116$).

41	36	12	18	28	23	19	8	7	16	11	14	18	14
34	6	30	11	1	11	4	32	23	45	115	37	29	71
39	23	21	37	20	12	13	135	49	32	64	40	77	97
97	85	10	27	7	48	35	61	79	63	16	80	108	20
52	82	50	64	59	39	9	16	78	35	66	122	89	110
44	28	65	22	59	23	31	44	21	9	45	168	73	76
118	84	85	96	78	73	91	47	32	20	23	21	24	44
21	28	9	13	46	18	13	24	16	13	23	36	7	14
30	14	18	20										

The MoM is used for estimating the parameters of the αIG and IG distributions. Note that the MoM estimators of the parameters α, λ and μ of the αIG distribution are given in Section 3. Here, function "fsolve" available in MATLAB2015b can be used to solve the system of equations given in Eq. (9). Also, the MoM estimators of the parameters λ and μ of the IG distribution are

$$\tilde{\lambda}_{\mathrm{MoM}} = \frac{\left[(1/n) \sum_{i=1}^{n} x_i \right]^3}{\left[(1/n) \sum_{i=1}^{n} x_i^2 - \left((1/n) \sum_{i=1}^{n} x_i \right)^2 \right]} \quad \text{and}$$

$$\tilde{\mu}_{\mathrm{MoM}} = (1/n) \sum_{i=1}^{n} x_i$$

respectively.

The MoM estimates of the parameters of the αIG and IG distributions along with the goodness-of-fit results based on them are given in Table 2.

As seen from Table 2, the KS, AD and RMSE values of the αIG distribution are smaller and R^2 value is greater than the corresponding values of the IG distribution. Hence, the αIG distribution provides a better fitting performance than the IG distribution. The fitting performance of the αIG and IG distributions are also shown in Fig. 2.

Table 2. Fitting results for the air pollution data.

	αIG Distribution					
$\hat{\alpha}_{\mathrm{MoM}}$	$\hat{\lambda}_{\mathrm{MoM}}$	$\hat{\mu}_{\mathrm{MoM}}$	KS	AD	RMSE	R^2
1.7245	125.1772	66.5585	0.0662	0.5016	0.0255	0.9926
	IG Distribution					
—	$\tilde{\lambda}_{\mathrm{MoM}}$	$\tilde{\mu}_{\mathrm{MoM}}$	KS	AD	RMSE	R^2
—	69.3113	42.1293	0.1007	2.7614	0.0511	0.9752

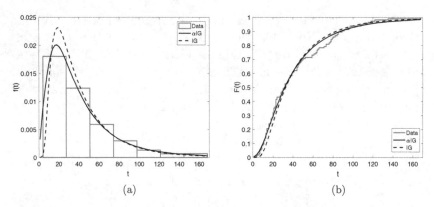

Fig. 2.   (a) PDF and (b) CDF fitting plots of the αIG and IG distributions.

5.   Conclusion

A new extension of the IG distribution, called αIG, is introduced in this study. Then, the MoM estimates of the parameters α, λ and μ of the αIG distribution are obtained. The αIG and IG distributions are used for modeling the air pollution dataset provided by Nadarajah (2008), and their fitting performances are compared by using the well-known goodness-of-fit statistics. The comparison result shows that the αIG distribution is preferable over the IG distribution for modeling the air pollution data.

References

Arslan, T. (2021a). An α-monotone generalized log-Moyal distribution with applications to environmental data. *Mathematics*, **9**(12): 1400.

Arslan, T. (2021b). An extension of the inverse Gaussian distribution. In *5th International Conference on Computational Mathematics and Engineering Sciences*, p. 194 (Van, Turkey).

Butler, R.W. (1998). Generalized inverse Gaussian distributions and their Wishart connections. *Scandinavian Journal of Statistics*, **25**(1): 69–75.

Chaudhry, M.A. and Zubair, S.M. (1994). Generalized incomplete gamma functions with applications. *Computational and Applied Mathematics*, **55**: 99–124.

Folks, J.L. and Chhikara, R.S. (1978). The inverse Gaussian distribution and its statistical application — A review. *Journal of the Royal Statistical Society Series B (Statistical Methodology)*, **40**(3): 263–289.

Iwase, K. and Hirano, K. (1990). Power inverse Gaussian distribution and its applications. *Japanese Journal of Applied Statistics*, **19**: 163–176.

Jones, M.C. (2020). On univariate slash distributions, continuous and discrete. *Annals of the Institute of Statistical Mathematics*, **72**: 645–657.

Junfeng, L.A.I., Danda, J.I. and Zaizai, Y. (2020). Extended inverse Gaussian distribution: Properties and application. *Journal of Shanghai Jiaotong University (Science)*, **25**(2): 193–200.

Lee, C., Famoye, F. and Alzaatreh, A.Y. (2013). Methods for generation families of univariate continuous distributions in the recent decades. *Wiley Interdisciplinary Reviews: Computational Statistics*, **5**: 219–238.

Lemonte, A.J. and Cordeiro, G.M. (2011). The exponentiated generalized inverse Gaussian distribution. *Statistics & Probability Letters*, **81**(4): 506–517.

Mathai, A.M. and Provost, S.B. (2011). The q-extended inverse Gaussian distribution. *Journal of Probability and Statistical Science*, **9**(1): 1–20.

Nadarajah, S. (2008). A truncated inverted beta distribution with application to air pollution data. *Stochastic Environmental Research and Risk Assessment*, **22**(2): 285–289.

Tweedie, M.C.K. (1945). Inverse statistical variaties. *Nature (London)*, **155**: 453.

Tweedie, M.C.K. (1957a). Statistical properties of inverse Gaussian distributions I. *The Annals of Mathematical Statistics*, **28**(2): 362–377.

Tweedie, M.C.K. (1957b). Statistical properties of inverse Gaussian distribution II. *The Annals of Mathematical Statistics*, **28**(3): 696–705.

Vasconcelos, J.C.S., Cordeiro, G.M., Ortega, E.M.M. and Araújo, E.G. (2019). The new-odd Log-Logistic generalized inverse Gaussian regression model. *Journal of Probability and Statistics*, Article ID: 857424: 1–13.

Chapter 11

Clustering Eurozone Countries According to Employee Contributions Before and After COVID-19

Hüseyin Ünözkan[*,†,§], **Nihan Potas**[‡,¶],
and Mehmet Yılmaz[†,‖]

[†]*Department of Statistics, Ankara University,
Ankara, Turkey*

[‡]*Department of Healthcare Management, Ankara Hacı
Bayram Veli University, Ankara, Turkey*

[§]*hunozkan@gmail.com*
[¶]*nihan_potas@hotmail.com*
[‖]*mehmetyilmaz@ankara.edu.tr*

Many researchers have tried to analyze economic situations with cluster analyses. In this study, we try to analyze the effects of coronavirus disease 2019 (COVID-19) on 29 Eurozone countries by changes of the clusters. The dataset contains species from the European Union formal data group, and they are gross domestic product (GDP) at current prices per hour worked, average annual hours worked per person employed, GDP at 2015 reference levels adjusted for the impact of terms of trade per person employed, real compensation per employee (deflator GDP: total economy) and real unit labor costs (total economy: ratio of compensation per employee to nominal GDP per person employed).

---

[*]Corresponding author.

We investigate the economic indicators of two different years independently. The cluster analysis for 2019 gives us two clusters for the 29 Eurozone countries. On the other hand, the cluster analysis with the same data group for 2020 gives three clusters. Some countries dissociate positively, while others are affected by COVID-19 negatively. The study shows that COVID-19 affected Eurozone countries in terms of certain European Union employee data group.

1.   Introduction

Cluster analysis is used commonly in economic datasets, especially to compare countries and identify similarities. The European Union is one of the oldest economic unions, and this community has kept economic records for a long time. This union established a European Commission and through this commission, a database was created. Economic data groups have been recorded continuously since 1962.

Coronovirus disease 2019 (COVID-19) emerged in December 2019 and has affected the economy of nearly every country. In the struggle against COVID-19, different states have adopted different precautions. Some of these precautions are directly or indirectly related to economies. Some precautions decrease efficiency of employees, whereas some others increase the efficiency. The duration of these precautions vary from state to state. Each precaution gives different result in each country.

Generally, exports, imports, inflation, gross domestic product (GDP), gross savings and exchange rates are used in the economic analysis of countries by cluster analysis. Cluster analysis is also widely used to make comparisons. This is usually done by looking at changes in clusters before and after major political events. For this, two separate clustering analyses are performed. By using the results of the analyses, we tried to determine how the countries have changed (Nastu *et al.*, 2019).

Although cluster analysis is commonly used in economic datasets, economic development level or economic growth level measures may vary in studies. To determine the economic situation in its member states, the European Union measures more than 500 different economic data annually. This dataset is grouped under major titles (European Union, 2021). Some titles measure the same situation with different approaches.

Cluster analysis is commonly used to compare regions after major economic or political decisions (Brauksa, 2013). Another important use of this analyze is determining how the effect of a major change differs from state to state. In this study, we investigate how the results of precautions for COVID-19 differ from country to country in the Eurozone. We try to obtain the differences with two different cluster analyses. First, we gain a cluster number with hierarchical cluster analysis, and later with k-means analysis, we obtain the clusters. These two steps repeat in both analyses for 2019 and 2020.

In a study, Göbel and Araújo (2020) investigated pre-crisis period macroeconomic indicators. They studied a dataset of 27 countries based on five variables to determine currency crises. Similar datasets were studied with cluster analyses to determine economic situation (Berg and Pattillo, 1999; Sarlin and Marghescu, 2011). In addition to these valuable studies, Göbel and Araújo (2020) try to determine early warning indicators by using cluster analysis and they examine these indicators working principles from state to state.

Although there are several studies on cluster analysis with economic indicators, we summarize here only some recent studies. Davidescu *et al.* (2015) try to study European countries and with his study he try to examine the outcome of eco-innovation for 27 countries in Europe during 2003–2013. The authors try to decide the location of the state of Romania during 2003–2013.

Vazquez and Sumner (2012) offer clustering for dividing countries into groups which indicate the grade of development. Their method is the hierarchical clustering with the Ward method which is based on the Euclidean square distance. The study examines 101 countries. The period considered by this study is 2005–2010.

Nastu *et al.* (2019) analyses the level of economic development of countries. They use 12 indicators in this study, and they extracted dataset from the World Bank database. They considered 60 countries from three continents for the year 2015.

The rest of this chapter is organized as follows. In the Methodology section, we summarize cluster analysis and introduce two different cluster analysis techniques which are used in this study. Analysis results are given in the Analysis section, and the Conclusion section concludes the study with comparisons.

2. Methodology

Clustering is a base composition of statistical analysis. Thus, clustering has a significant role in defining a statistical situation. Although many different techniques for cluster analysis have been created, some of them are more popular in scientific studies.

Distances (dissimilarities) and similarities of clusters are the basis for constructing clustering algorithms. In respect of quantitative data features, distance is used to determine the relationship. This leave its place to similarity when analyzing with qualitative data features (Xu and Wunsch, 2005; Xu and Tian, 2015).

The main idea of classic clustering algorithms is to reference the center of data points as the center of the corresponding cluster. k-means is the most famous of this kind of clustering algorithms. The basic idea of k-means is to recalculate the centers of clusters which are represented by the centers of data points, and this iterative process will be continued until the defined criteria for convergence is met (MacQueen, 1967).

A popular clustering algorithm is the k-means approach by Newman (2006), in which the researcher assumes that countries have similar dynamics for a determined period of time to later experience a similar economic state. In the end, countries with more dissimilar macroeconomic appearance are required to be in a different cluster and not with the same cluster for economic situation. Thus, the researcher concluded that the biggest distance means biggest differences rather than small distances.

Another famous clustering algorithm is the construction of a hierarchical relationship among data in order to cluster (Johnson, 1967). Assume that every data point stands for a unique cluster at the beginning, and later, the closest two clusters form a new cluster until there is only one cluster left.

Generally, typical hierarchical clustering realizes the results by constructing a feature tree of clustering, in which each node stands for a subcluster. Clustering feature tree dynamically grows once a new data point occurs (Zhang *et al.*, 1996).

In hierarchical clustering, Euclidean distance usage is very popular (Göbel and Araújo, 2020). This distance, as described by Gan *et al.* (2007), between country n and country z for a particular quarter t is defined as

$$d_t\left(v_{n,t}, v_{z,t}\right) = \left[\sum_{t=1}^{l}\sum_{i=1}^{l}\left(p_{i,n,t} - p_{i,z,t}\right)^2\right]^{\frac{1}{2}}$$

$$= \left[\left(v_{n,t} - v_{z,t}\right)\left(v_{n,t} - v_{z,t}\right)^T\right]^{\frac{1}{2}} \tag{1}$$

where $p_{i,n,t}$ and $p_{i,z,t}$ are the percentiles of the ith variable of countries n and z, respectively, for the quarter t.

Once the distance in Eq. (1) is generalized and the distance matrix is obtained, the new equation can be gained as follows:

$$d_{\Delta t}\left(v_{n,\Delta_t}, v_{z,\Delta_t}\right) = \left[\sum_{t=1}^{l}\sum_{i=1}^{l}\left(p_{i,n,t} - p_{i,z,t}\right)^2\right]^{\frac{1}{2}}$$

$$= \left[\left(v_{n,\Delta_t} - v_{z,\Delta_t}\right)\left(v_{n,\Delta_t} - v_{z,\Delta_t}\right)^T\right]^{\frac{1}{2}} \tag{2}$$

In the literature, there are many other distance measurement methods for hierarchical cluster analysis (Göbel and Araújo, 2020; Mantegna, 1999; Prekopcsák and Lemire, 2012).

In this study, at first we use hierarchical cluster analysis to gain appropriate cluster numbers, later we use k-means cluster analysis to gain clusters for the Eurozone countries.

3. Analysis

The dataset under study contains yearly GDP at current prices per hour worked, average annual hours worked per person employed, GDP at 2015 reference levels adjusted for the impact of terms of trade per person employed, real compensation per employee (deflator GDP: total economy) and real unit labor costs (total economy: ratio

Table 1.   Descriptive statistics of the variables.

Variable	Definition
Variable-1	GDP at current prices per hour worked
Variable-2	Average annual hours worked per person employed
Variable-3	GDP at 2015 reference levels adjusted for the impact of terms of trade per person employed
Variable-4	Real compensation per employee, deflator GDP:total economy
Variable-5	Real unit labor costs:total economy (ratio of compensation per employee to nominal GDP per person employed)

of compensation per employee to nominal GDP per person employed) for the years 2019 and 2020. The dataset is obtained from the European Union database (European Union, 2021). The variables are listed in Table 1.

Figure 1 represents the dendogram of the 2019 dataset.

The dendogram offers us two-cluster modeling for appropriate analysis. According to Ward method with Euclidean distance, we continue with two clusters in k-means cluster analysis.

The test results of k-means cluster analysis with two clusters is shown in Tables 2–4. In Table 2, the distance between the clusters is shown.

In Table 3, an ANOVA table gives the analysis of distances of the means from both clusters. According to Table 3, there are significant differences between clusters for each variable.

Based on Table 4, we conclude that in 2019, with the five variables we use, there are two clusters and in the first cluster there are 12 countries and the second cluster has 17 countries.

Figure 2 represents the dendogram of the 2020 dataset.

The dendogram offers us three-cluster modeling for appropriate analysis. According to Ward method with Euclidean distance, we continue with three clusters in k-means cluster analysis.

The test results of k-means cluster analysis with three clusters is shown in Tables 5–7. In Table 5, the distances from clusters are shown. The distances between clusters 2 and 3 is the smallest value.

In Table 6, an ANOVA table gives the analysis of distances about means from both clusters. According to Table 6, there are significant differences between clusters for each variable.

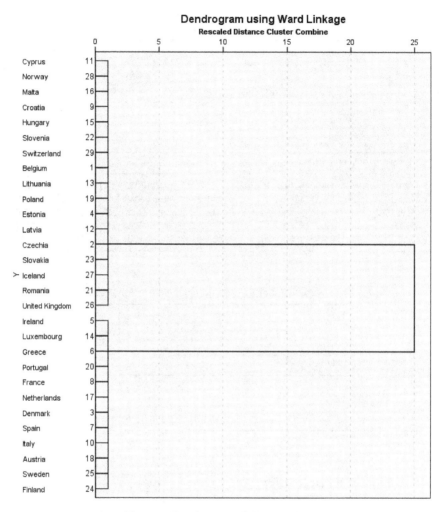

Fig. 1.   Dendogram of 2019 dataset.

Based on Table 6, we conclude that in 2020 with the five variables we use, there are three clusters and in the first cluster there are three countries, the second cluster has 19 countries and the third cluster has seven countries.

Based on Table 7, we may conclude that some states positively dissociate with gains from precautions, whereas some states negatively dissociate.

Table 2.　Distances between final cluster centers.

Cluster	1	2
1		1934529169.292
2	1934529169.292	

Table 3.　ANOVA of k-means for 2019.

	Cluster Mean Square	df	Error Mean Square	df	F	Sig.
VAR2019_1	9583595890957013000	1	55832266815796512	27	171.650	0.000
VAR2019_2	53013308818420.64	1	3842368404359.315	27	13.797	0.001
VAR2019_3	8268693608328434700	1	16207632660301860	27	510.173	0.000
VAR2019_4	1581488215483903490	1	2235338687709673	27	707.494	0.000
VAR2019_5	6892039402795532300	1	273890107633423.160	27	25163.521	0.000

Table 4.　Number of cases in each cluster for 2019.

Cluster	1	12.000
	2	17.000
Valid		29.000
Missing		0.000

Table 5.　Distances between final cluster centers.

Cluster	1	2	3
1		1054779160.757	1074072916.087
2	1054779160.757		256848232.531
3	1074072916.087	256848232.531	

Table 6.　ANOVA of k-means for 2020.

	Cluster Mean Square	df	Error Mean Square	df	F	Sig.
VAR2020_1	65990352536915.766	2	1515178311412.45	26	43.553	.000
VAR2020_2	14927755561936369410	2	211661404365057	26	7052.658	.000
VAR2020_3	39598557471922.305	2	5523591824655,234	26	7.169	.003
VAR2020_4	143268117686115696	2	2996732766646054	26	47.808	.000
VAR2020_5	26200689922663688	2	2802012879201782	26	9.351	.001

Table 7.   Number of cases in each cluster for 2020.

Cluster	1	3.000
	2	19.000
	3	7.000
Valid		29.000
Missing		0.000

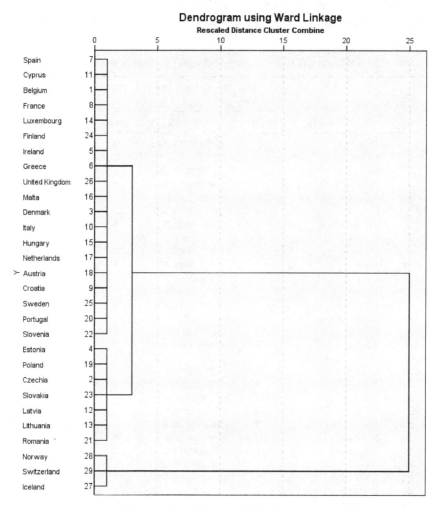

Fig. 2.   Dendogram of 2020 dataset.

4. Conclusion

In this study, we aimed to determine the effects of precautions for COVID-19 on countries' economies in the Eurozone. We proposed cluster analysis because we needed to obtain the differences of locations for countries after COVID-19. In Table 8, we can see the locations of countries before and after the effects of COVID-19.

Based on Table 8, we conclude that Norway, Switzerland and Iceland dissociate positively with their employee precaution performances. After these three countries, the United Kingdom, Slovenia,

Table 8.   Performance measures of countries' precaution results.

Case Number	Nationality	Cluster Membership for 2019		Cluster Membership for 2020	
		Cluster	Distance	Cluster	Distance
1	**Belgium**	2	**108243730.354**	2	**56305626.136**
2	Czechia	2	87015676.278	3	99433620.021
3	Denmark	1	213233024.154	2	42394276.273
4	Estonia	2	96580128.349	3	65998611.356
5	Ireland	1	649525099.863	2	139294709.165
6	Greece	1	587062425.065	2	73563222.265
7	Spain	1	220158813.949	2	26363088.386
8	France	1	92099944.232	2	57923183.746
9	**Croatia**	2	**76404247.342**	2	**26956052.663**
10	Italy	1	170306421.443	2	46535320.764
11	Cyprus	2	87533122.952	2	26317445.114
12	Latvia	2	141575938.910	3	57602716.579
13	Lithuania	2	140298217.341	3	64453228.316
14	Luxembourg	1	589067271.596	2	47988747.301
15	**Hungary**	2	**94771818.758**	2	**51752471.958**
16	**Malta**	2	**69008832.288**	2	**117248510.271**
17	Netherlands	1	59126938.776	2	15067944.871
18	Austria	1	25853155.358	2	15838639.748
19	Poland	2	137142960.933	3	67527209.819
20	Portugal	1	511089074.670	2	49460271.829
21	Romania	2	277521138.144	3	164647339.170
22	**Slovenia**	2	**84660103.246**	2	**91639180.144**
23	Slovakia	2	80200853.538	3	92549514.687
24	Finland	1	93984649.123	2	105428112.141
25	Sweden	1	37181439.285	2	20629911.876
26	**United Kingdom**	2	**518498988.133**	2	**58630510.292**
27	**Iceland**	2	**30635220.869**	1	**76661986.196**
28	**Norway**	2	**75851838.984**	1	**39856195.364**
29	**Switzerland**	2	**108820699.777**	1	**67772450.001**

Note: Countries successfully performing employee precautions are in bold.

Malta, Hungary, Croatia and Belgium perform a successful employee precaution models.

In the future, the effects of COVID-19 on economies needs to be studied further. The effects of COVID-19 on economies may be seen in some years or decades in some countries or may be in some economic or political unions. According to our study, we propose that the employee precaution performances of countries differ among the Eurozone countries.

References

Berg, A. and Pattillo, C. (1999). What caused the asian crises: An early warning system approach. *Econ Notes Banca Monte dei Paschi di Siena SpA*, **28**(3): 285–334.

Brauksa, I. (2013). Use of cluster analysis in exploring economic indicator differences among regions: The case of Latvia. *Journal of Economics, Business and Management*, **1**(1): 42–45.

Davidescu, A., Paul, A., Gogonea R.M. and Zaharia M. (2015). Evaluating Romanian eco-innovation performances in European context. *Sustainability*, **7**(9): 12723–12757.

European Commission (2021). Eurostat your key to European statistics. Accessed 1 July 2021. https://ec.europa.eu/eurostat/.

Gan, G., Ma, C. and Wu, J. (2007). Data clustering: Theory, algorithms, and applications. *Society for Industrial and Applied Mathematics*, 3–17.

Göbel, M. and Araújo, T. (2020). Indicators of economic crises: A data-driven clustering approach. *Applied Network Science*, **5**(44): 1–20.

Johnson, S. (1967). Hierarchical clustering schemes. *Psychometrik*, **32**: 241–254.

MacQueen, J. (1967). Some methods for classification and analysis of multivariate observations. *Proceedings of the Fifth Berkeley Symposium on Mathematical Statistics and Probability*, **32**: 241–254.

Mantegna, R.N. (1999). Hierarchical structure in financial markets. *European Physical Journal B*, **11**: 193–197.

Nastu, A., Stancu, S. and Dumitrache, A. (2019). Characterizing the level of economic development of countries. In *2019 Proceedings of the 13th International Conference on Applied Statistics*, (pp. 343–354) Bucharest.

Newman, M. (2006). Modularity and community structure in networks. *Proceedings of the National Academy of Sciences*, **103**(23): 8577–8582.

Prekopcsák, Z. and Lemire, D. (2012). Time series classification by class-specific Mahalanobis distance measures. *Intelligent Systems in Accounting, Finance and Management*, **6**(3): 185–200.

Sarlin, P. and Marghescu, D. (2011). Visual predictions of currency crises using self-organizing maps. *Advances in Data Analysis and Classification*, **18**(1): 15–38.

Vazquez, S.T. and Sumner, A. (2012). Beyond low and middle income countries: What if there were five clusters of developing countries? Poverty and inequality research cluster IDS Working Papers.

Xu, R. and Wunsch, D. (2005). Survey of clustering algorithms. *IEEE Transactions on Neural Networks*, **16**: 645–678.

Xu, D. and Tian, Y. (2015). A comprehensive survey of clustering algorithms. *Annals of Data Science*, **16**: 645–678.

Zhang, T., Ramakrishnan, R. and Livny, M. (1996). An efficient data clustering method for very large databases. *ACM SIGMOD Record*, **25**: 103–104.

Chapter 12

Criteria for Best Architecture Selection in Artificial Neural Networks

Çağatay Bal*,† and Serdar Demir‡

*Department of Statistics, Muğla Sıtkı Kocman University,
Muğla, Turkey*

†*cagataybal@mu.edu.tr*
‡*serdardemir@ mu.edu.tr*

Architecture selection in artificial neural networks is a critical process which determines a satisfactory neural network model(s) that will lead to the most accurate results. The architecture that minimizes the difference between the target values of the neural network and the predictions produced by the model represents the best forecasts, namely the most appropriate model. In the literature, there are many common criteria for measuring model performance. In addition, some modified criteria, called weighted criteria, are suggested by combining the common criteria. In this study, the performances of the criteria available in the literature are compared by using both simulated and real-world datasets. We used three different exchange rate time series, four simulated time series with different structures and three well-known real-world datasets. The results show that the performances of the unweighted criteria vary depending on the data structure. However, the weighted criteria have performances as good as the popular criteria or better.

---

*Corresponding author.

1.　Introduction

As well known, artificial neural networks (ANNs) are powerful heuristic approaches founded by mimicking the neural communication structure of the human brain. The term *architecture* is one of the hyperparameters of ANNs, and it describes the structure of the network. Learning ability is the most important characteristic feature of ANNs, and it is made possible by *backpropagation learning algorithm* which was introduced by Werbos (1974) and successfully implemented in ANNs by Vogl *et al.* (1988). Learning algorithm is a hyperparameter of ANNs which plays a key role in successful model fitting and forecasting. The adaptation of nonlinearity in ANNs could be achieved by using *activation functions*. These functions have the ability to shape and adapt the data in order to map the relationship within the observations and acquire the desired results from different types of networks for various tasks.

Using ANNs for various forecasting tasks has been done successfully over the years. The reason is that ANNs are also called "Universal Approximators," which means ANNs with at least one hidden layer can approximate any continues function (Cybenko, 1989; Hornik, 1991).

The main focus is out-of-sample predictions on forecasting which makes it different from fitting or modeling the data. In detail, a model can be acquired by using the data and the performance of the model can be tested by in-sample observations which have already been used to create the model. However, it might not mean that the well-trained model can produce successful forecasts every time (Hyndman and Koehler, 2006). The expectation from the well-trained model is to predict the unseen out-of-sample observations accurately. But most of the time, these expectations don't match the need, and it becomes hard to calculate the forecasting performance of the model to determine whether or not to use the model for forecasting.

The first ANNs are called "Perceptron" (McCulloch and Pitts, 1943) and only able to handle linear problems as a linear classifier. Perceptron's architecture has only a single layer and a basic type of learning rule with a step function as the activation in its network. These properties are the reasons why it can only be used as linear classifier. But after late 1980s, with the development of new components such as learning algorithms, activation functions,

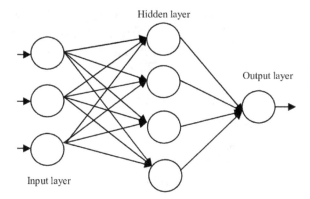

Fig. 1.   Architecture of ANNs.

improved architectures and specialized networks, "Multi-Layer Perceptron" (MLP), mainly called ANNs, has become very successful and is recognized by numerous researchers from all around the world. The architecture of MLP can be seen in Fig. 1.

The goal of forecasting with ANNs is obtaining a network with better out-of-sample performance than in-sample performance, implying that the best network is not always well trained but performs well at out-of-sample forecasting. Networks with perfect in-sample fit and poor out-of-sample performance cause a situation called *overfitting* in this field, and to train a proper network comes with various strategies to avoid overfitting such as cross-validation (Krogh and Vedelsby, 1995), k-fold cross-validation (Fushiki, 2011), noise injection (Zur *et al.*, 2009), regularization (Zaremba *et al.*, 2014) and the classic way of dividing data into training, testing and sometime validation set.

As the classic and the effective way of avoiding overfitting, splitting the data into training and testing is used in this study. ANNs are trained by using a training set and the selected model is tested for out-of-sample performance using a test set. Validation set partition is usually not preferred in time series forecasting with ANNs (Zhang *et al.*, 1998). Networks with the best test set performances are determined as appropriate models.

ANNs have the model selection dilemma due to random initialization of weights at the beginning of training process. Random initialization of weights can be seen as an advantageous property of

ANNs because this enables alternative solutions in the solution space (Bal and Demir, 2017). On the other hand, random initialization of weights will lead to a problem of having different architectures after training for the same data. This mostly causes inconsistencies in the interpretation of the results of the network. For this reason, the training process often consists of training many networks simultaneously and selecting the best performed among obtained ones.

2. Performance Measures of Forecasting

In time series analysis, let the dependent variable, also named target observations in ANNs, be y_t and the explanatory variables are the past time lags of y_t which can be seen as follows:

$$y_t = y_{t-1} + y_{t-2} + \cdots + y_{t-k} \tag{1}$$

The strategy to achieve a powerful network is finding the relationship between $y_{t-1}, y_{t-2}, \ldots, y_{t-k}$ and y_t with a good test set performance. To calculate the performance of network, an error term must be obtained by using the forecasts $\breve{y}_t$ and observed values y_t. Calculating test set performance of network can be done by performance criteria based on the error term (e_t) based on the distance between y_t and $\breve{y}_t$. The list of error types are shown in Table 1.

Table 1.   List of error types.

Error Type	Formula
Error	$e_t = (y_t - \breve{y}_t)$
Percentage error	$p_t = \frac{(y_t - \breve{y}_t)}{y_t}$
Symmetric error	$s_t = \frac{|y_t - \breve{y}_t|}{(y_t + \breve{y}_t)}$
	$s_t^* = mean_{k=1,i-1}|y_k - \bar{y}_{i-1}|$
Relative error	$r_t = \frac{|y_t - \breve{y}_t|}{(y_t - f_t^*)}$ $f_t^* = y_{t-l}$
Scaled error	$sc_t = \frac{|y_t - \breve{y}_t|}{\frac{1}{n-1}\sum_{t=2}^{n}|y_t - y_{t-l}|}$

2.1. *Absolute-error-based criteria*

The criteria based on e_t are always on the same scale as the data. These criteria cannot be used to compare time series with different scales. In such a case, the preprocessing of data may be required to obtain the same scale for each dataset. The performance measures based on e_t are shown in Table 2.

2.2. *Percentage-error-based criteria*

The criteria based on percentage error (p_t) have the advantageous of being scale-independent across different scaled datasets. However, the major disadvantage of p_t is division by zero when y_t is equal or very close to zero. This can cause undefined or infinite results when measuring the performance. The performance measures based on p_t are shown in Table 3.

Table 2. Performance measures based on e_t.

Performance Measures	Formulas
Mean absolute error	$\mathrm{MAE} = \mathrm{mean}_{i=1,n}\lvert e_i\rvert$
Median absolute error	$\mathrm{MdAE} = \mathrm{median}_{i=1,n}\lvert e_i\rvert$
Geometric mean absolute error	$\mathrm{GMAE} = g\mathrm{mean}_{i=1,n}\lvert e_i\rvert$
Mean square error	$\mathrm{MSE} = \mathrm{mean}_{i=1,n}(e_i^2)$
Root mean square error	$\mathrm{RMSE} = \sqrt{\mathrm{mean}_{i=1,n}(e_i^2)}$
Fourth root mean quadrupled error	$\mathrm{R4MS4E} = \sqrt[4]{\mathrm{mean}_{i=1,n}(e_i^4)}$

Table 3. Performance measures based on p_t.

Performance Measures	Formulas
Mean absolute percentage error	$\mathrm{MAPE} = \mathrm{mean}_{i=1,n}\lvert p_i\rvert$
Median absolute percentage error	$\mathrm{MdAPE} = \mathrm{median}_{i=1,n}\lvert p_i\rvert$
Root mean square percentage error	$\mathrm{RMSPE} = \sqrt{\mathrm{mean}_{i=1,n}(p_i^2)}$
Root median square percentage error	$\mathrm{RMdSPE} = \sqrt{\mathrm{median}_{i=1,n}(p_i^2)}$

Another issue with p_t is that it should be based on quantities such as exchange rates and number of products sold because it doesn't make sense to represent performance with percentage if the data does not include quantities such as temperatures in Celsius or Fahrenheit (Hyndman and Koehler, 2006). The main reason is that percentage should not change when the scale of different datasets is changed, and this is only eligible for quantity-based observations. It should also be noted that the criteria based on p_t are not symmetric, which simply means that results may differ with predictions being higher or lower than actual observed values.

2.3. *Symmetric-error-based criteria*

Symmetric error (s_t) is a type of percentage error which modified to add symmetry to the p_t. However, as stated by Koehler (2001), despite its name, this error is actually not symmetric. The criteria based on s_t share the similar disadvantage as p_t of division by zero when y_t and $\breve{y}_t$ are close to zero. To overcome this situation, a modified version (s_t^*) can be used. The performance measures based on s_t are shown in Table 4.

2.4. *Relative-error-based criteria*

The main feature of the criteria based on relative error (r_t) is the use of a benchmarking approach for measuring the relativity between the errors produced by the chosen benchmark method and f_t^* errors (e.g. usually naive method). The criteria based on r_t share the similar disadvantage as p_t and s_t of division by zero when y_t and f_t^* are close to zero. The performance measures based on r_t are shown in Table 5.

Table 4.   Performance measures based on s_t.

Performance Measures	Formulas
Symmetric mean absolute percentage error	$\text{SMAPE} = \text{mean}_{i=1,n}(s_i)$
modified symmetric mean absolute percentage error	$\text{MSMAPE} = \text{mean}_{i=1,n}(s_i/s_t^*)$
Symmetric median absolute percentage error	$\text{SMdAPE} = \text{median}_{i=1,n}(s_i)$

Table 5. Performance measures based on r_t.

Performance Measures	Formulas
Mean relative absolute error	$\text{MRAE} = \text{mean}_{i=1,n}|r_i|$
Median relative absolute error	$\text{MdRAE} = \text{median}_{i=1,n}|r_i|$
Geometric mean relative absolute error	$\text{GMRAE} = g\text{mean}_{i=1,n}|r_i|$

Table 6. Performance measures based on sc_t.

Performance Measures	Formulas
Mean absolute scaled error	$\text{MASE} = \text{mean}_{i=1,n}|sc_i|$
Root mean square scaled error	$\text{RMSSE} = \sqrt{\text{mean}_{i=1,n}(sc_i^2)}$

2.5. *Scaled-error-based criteria*

Scaled error (sc_t) was proposed by Hyndman and Koehler (2006) to enable the comparison of forecasting accuracy from scale-independent perspective. The criteria based on sc_t share a similar behavior with p_t and r_t but though they have certain pros such as being scale-independent unlike p_t, the disadvantage of division by zero still exist but with less likelihood to occur than p_t. Also, sc_t uses in-sample MAE from the naive method as a benchmark similar to r_t but the relative errors have a undefined mean and infinite variance, which can only be used when there are several forecasts on the same series, so it cannot be used to measure out-of-sample forecast accuracy at a single forecast horizon (Hyndman and Koehler, 2006). The performance measures based on sc_t are shown in Table 6.

2.6. *Weighted criteria*

The criteria based on weighted errors are those formed by using the other performance criteria together. Weighted errors are generated by multiplying the performance criteria selected according to the user's preference with the coefficients a_i created based on certain strategies or preferences. The point here is that the sum of the coefficients must be equal to one, i.e. $\sum a_i = 1$.

The weighted information criterion (WIC) developed for forecasting time series is given as follows (Egrioglu *et al.*, 2008):

$$\text{WIC} = 0.2 \times (\text{RMSE} + \text{MAPE} + \text{MDA} + (1 - \text{DA}))$$
$$+ 0.1 \times (\text{AIC} + \text{BIC}) \tag{2}$$

WIC is a weighted criterion which consists of RMSE, MAPE, DA, MDA, AIC and BIC with constant coefficients. The AIC and BIC coefficients were set as 0.1 because they were considered of less importance for the reason of their penalizing structure. Other criteria were considered more important, and their coefficients were determined as 0.2.

The adaptive version of WIC has been developed and named adaptive WIC which is given as follows (Aladağ *et al.*, 2010):

$$\text{AWIC} = a_1 \times \text{RMSE} + a_2 \times \text{MAPE} + a_3 \times \text{MDA}$$
$$+ a_4 \times (1 - \text{DA}) + 0.1 \times (\text{AIC} + \text{BIC}) \tag{3}$$

AWIC is an information criterion in which the weights are determined in a data-specific manner through an optimization process and can also have some fixed weight coefficients. The coefficient of each criterion is calculated in accordance with the objective function with an approach determined by the user. The strategy used is to calculate AWIC with coefficients that maximize the correlation between these two test sets by forming the test set in two parts (Aladağ *et al.*, 2010).

Another weighted performance criterion named extended weighted performance criterion (EWPC) is given as follows (Cagatay Bal *et al.*, 2016):

$$\text{EWPC} = 1/19 \times (\text{MSE} + \text{RMSE} + \text{R4MS4E} + \text{MAPE} + \text{MAE}$$
$$+ \text{GMAE} + \text{MdAE} + \text{MdAPE} + \text{NS} + \text{MRAE} + \text{MdRAE}$$
$$+ \text{GMRAE} + \text{RMSPE} + \text{RMdSPE} + \text{SMAPE}$$
$$+ \text{SMdAPE} + \text{MSMAPE} + \text{MASE} + \text{RMSSE}) \tag{4}$$

EWPC is a performance criterion with constant coefficients, which consists of 19 different performance measures and has the characteristics of each. As it can be seen, the EWPC criterion does not include the in-sample information criteria, unlike AIC and BIC which are based on punishment. The disadvantage of having the punishment term causes these criteria to ignore alternative neural network architectures (Cagatay Bal *et al.*, 2016).

2.7. *Other criteria*

Akaike information criterion (AIC) is a well-known criterion used commonly for time series forecasting:

$$\text{AIC} = \log\left(\frac{\sum_{t=1}^{N}(y_t - \breve{y}_t)^2}{N}\right) + \frac{2 * m}{N} \tag{5}$$

Here, N is the dataset size and m is the number of weights in the neural network. The first term of AIC equation includes the maximum likelihood estimation of the variance of residues. Thus, the first term implies the goodness of fit of the model with the data, and the second term implements punishment so that the model does not encounter overfitting. Theoretically, AIC is a reasonable criterion that includes balanced model fit and stinginess of the model. However, AIC is designed as an in-sample "information" criterion which is obviously not a good choice to use for the measurement of out-of-sample performance of a model or network. The optimal model is obtained when the AIC value reaches the minimum.

Second, Bayesian information criterion created by the differentiation of AIC is presented:

$$\text{BIC} = \log\left(\frac{\sum_{t=1}^{N}(y_t - \breve{y}_t)^2}{N}\right) + \frac{m * \log(N)}{N} \tag{6}$$

This is also a popular criterion and very similar to AIC. But its punishment term punishes more for complex models than AIC in cases where N is greater than seven. The appropriateness of the punishment term also requires a discussion on the number of parameters. The constant m makes the criterion more appropriate in terms of flexibility. But there is no research in the literature regarding the selection of an appropriate m value for nonlinear model selection. BIC is a consistent criterion for autoregressive (AR) models (Rissanen, 1980). In real-life applications, BIC is preferred because of having more reliable results than the AIC.

Third, Nash–Sutcliffe efficiency coefficient (NS) is presented:

$$\text{NS} = 1 - \frac{\sum_{t=1}^{n}(y_t - \check{y}_t)^2}{\sum_{t=1}^{n}(y_t - \bar{y})^2} \tag{7}$$

NS (Nash and Sutcliffe, 1970) is obtained by dividing the distance between observations and predictions by the variation of observations. The performance increases as NS approaches zero.

Fourth, direction accuracy (DA) is as follows:

$$\text{DA} = \frac{1}{N}\sum_{t=1}^{N} a_t, a_t = \left\{ \begin{matrix} 1 & \text{if } (y_{t+1} - y_t)(\check{y}_{t+1} - y_t) > 0 \\ 0 & \text{other situations} \end{matrix} \right\} \tag{8}$$

DA is a performance criterion which measures whether the estimation and observation values are in same direction. For the predicted values increasing or decreasing in the same direction, the observed values at times t and $t+1$ take the value of one, and zero for different directions and consists of the means of these values. As DA gets closer to one, equally good estimation values are obtained.

Finally, modified direction accuracy (MDA) is given as follows:

$$\text{MDA} = \frac{\sum_{t=1}^{N-1} D_t}{N-1} \tag{9}$$

$$D_t = (A_t - F_t)^2$$

$$A_t = 1, \quad y_{t+1} - y_t \le 0$$
$$A_t = 0, \quad y_{t+1} - y_t > 0$$
$$F_t = 1, \quad \check{y}_{t+1} - \check{y}_t \le 0$$
$$F_t = 0, \quad \check{y}_{t+1} - \check{y}_t > 0 \tag{10}$$

MDA is inspired by DA but more robust than DA.

All these performance criteria have certain advantages and disadvantages. The disadvantage of those criteria that use the average operator is that they are highly influenced by extreme values. The disadvantage of those criteria that use the median operator is that calculations get hard as the dataset grows. AIC and BIC penalize large networks as they have many parameters that need to be determined, such as number of layers, number of neurons, learning algorithm and activation function.

3.   Applications

In order to evaluate the application performances of the criteria discussed in this study, both real-world datasets and simulated datasets are used. For all applications, the modeling process is basically carried out in three stages.

- In the first stage, delay matrices are obtained from the dataset containing input and target values to be used in the neural network.
- In the second stage, 144 networks are created for each dataset. The number of neurons in the input and hidden layers ranges from 1 to 12, which is generally considered sufficient (Aladağ *et al.*, 2010). Then, the neural networks are trained and the performance criteria values are calculated for each network. The best performing network among the 144 networks is chosen for each dataset.
- In the last stage, the procedure in the second stage is repeated 1000 times. So, 1000 networks with the best performance are determined for each dataset. Finally, the values of all performance criteria, one-step ahead forecasts, correlation coefficients for test sets, weights of trained neural networks and test set of each network are obtained.

Levenberg–Marquardt backpropagation algorithm is used during the training phase. This algorithm is an easy-to-use algorithm with very fast convergence speed. The tangent sigmoidal function is used as the activation function in the hidden layer. In the output layer, the linear function is used in order not to compress the results in a certain range for this type of forecasting task.

3.1.   *Applications for simulated data*

The first simulated dataset is generated from Mackey–Glass (MG) time-delay differential equation (Glass and Mackey, 1979), which is a chaotic time series with no clearly defined period and also does not converge or diverge, making it suitable as a benchmark dataset. The second simulated dataset is generated from seasonal autoregressive (SAR1) process. The third dataset is generated from trigonometric sinus function. Finally, the fourth dataset is generated from a formula with exponential properties. Each of the simulated time series data

Table 7.   Simulated datasets and formulas.

Simulated Dataset	Formula
Mackey–Glass (MG)	$\frac{dx}{dt} = \beta\frac{x\tau}{1} + x\tau n - \gamma x,\quad \gamma\beta n >$ where β, γ, τ, n are real numbers, and $x\tau$ represents the value of the variable x at time $(t-\tau)$.
Seasonal autoregressive (SAR1)	$\text{SARIMA}(1,0,0)(1,1,0)_{12}$
Sinus	$\sin\left(2\pi x_t^3\right)$ where x_t is randomly generated from *uniform* distribution.
Exponential	$x_{t+1} = \frac{2x_t}{1+x_t^2} + e_t$ where e_t is randomly generated from *uniform* distribution and $x_t = 1$.

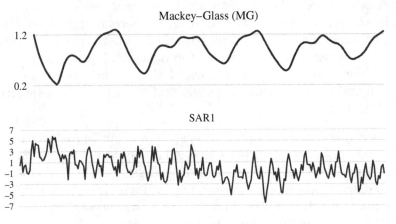

Fig. 2.   Graphs of MG and SAR1 time series.

consists of 256 data points. The formulas used to create the simulated datasets are given in Table 7. Scatter plots of the simulated time series data are given in Figs. 2 and 3.

The results of MG time series are given in Table 8.

From Table 8, for the MG dataset, it can be said that all criteria have similarly high concordance and error performance. MdAE has the best forecasting performance with the proximity of 98.58208%.

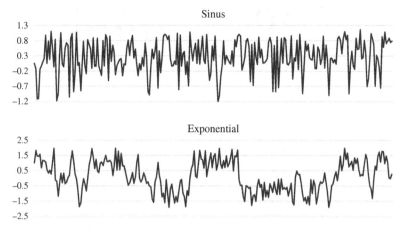

Fig. 3. Graphs of sinus and exponential time series.

It is seen that 1-1-1 and 1-3-1 architectures are predominantly chosen by many performance criteria for MG.

The coefficients of variation for the errors of the 1000 networks are computed for each performance criterion. Apart from WIC, 1-2-1 and 1-3-1 architectures have the best performance over 1000 repetitions. RMSE, RMSPE and NS have higher coefficients of variation and MAPE and SMAPE have lower coefficients of variation among other criteria.

Figure 4 shows the line graphs of the minimum error values obtained for 1000 repetitions for all the performance criteria of MG (error values are rescaled to 0–1 range for comparison). It is seen that MdAE, MAPE, SMAPE, MRAE and MASE have different characteristics compared to the others. Although these criteria have lower coefficients of variation, it can be seen that they do not perform well in selecting architectures with low error values. It is seen that the EWPC, WIC, RMSE, RMSPE and NS criteria have very similar characteristics with each other and perform better in selecting architectures with low error values among other criteria. However, it is necessary to draw attention to the fact that the lowest error frequency decreases considerably after 0.2 for EWPC.

The results of SAR1 time series are given in Table 9. From Table 9, the WIC criterion has the highest concordance performance with very high average correlation for the SAR1 and MASE has the highest error performance with very low average absolute error. MRAE has

Table 8. MG results.

	Average of Correlation Values Between Test Set and Estimates	Mean Absolute Errors Between Test Set and Estimates	Average of One-Step-Ahead Forecasts, Actual Value = 1.2666	Percentage Proximity to One-Step-Ahead Real Value	Most Selected Architecture and Number of Selection
			MG		
EWPC	1.0000	0.0018	1.2424	98.09%	1-2-1/487
WIC	1.0000	0.0018	1.2426	98.10%	1-1-1/1000
RMSE	1.0000	0.0018	1.2424	98.09%	1-1-1/506
MdAE	0.9999	0.0018	**1.2486**	**98.58%**	1-3-1/886
MAPE	0.9999	0.0017	1.2459	98.36%	1-3-1/428
RMSPE	1.0000	0.0018	1.2424	98.09%	1-1-1/504
SMAPE	0.9999	0.0017	1.2459	98.36%	1-3-1/428
MRAE	0.9999	0.0017	1.2458	98.36%	1-3-1/424
MASE	0.9999	0.0017	1.2458	98.36%	1-3-1/424
NS	1.0000	0.0018	1.2424	98.09%	1-1-1/506

Table 8. (*Continued*)

MG

	Minimum Error Value	Architecture of the Minimum Error Value	Maximum Error Value	Mean and Median of Error Values Set to 0–1 Interval		Standard Deviation of Error Values Set to 0–1 Interval	Coefficient of Variation of Error Values Set to 0–1 Interval
EWPC	0.0101	1-3-1	0.0107	0.6371	1.0000	0.3760	59.0229
WIC	−2.3716	1-1-1	−2.3716	0.6204	0.7610	0.3445	55.5296
RMSE	0.0024	1-2-1	0.0025	0.5221	1.0000	0.4997	95.7136
MdAE	0.0004	1-3-1	0.0011	0.4701	0.4241	0.1756	37.3594
MAPE	0.0014	1-3-1	0.0017	0.6949	0.6673	0.1222	17.5870
RMSPE	0.0021	1-3-1	0.0022	0.5221	1.0000	0.4997	95.7042
SMAPE	0.0014	1-3-1	0.0017	0.6973	0.6670	0.1227	17.6002
MRAE	0.0015	1-3-1	0.0018	0.7237	0.6808	0.1340	18.5152
MASE	0.0637	1-3-1	0.0748	0.7284	0.6759	0.1371	18.8178
NS	0.0002	1-2-1	0.0002	0.5221	1.0000	0.4997	95.7139

Note: Bold entries are min-max values of the respective column.

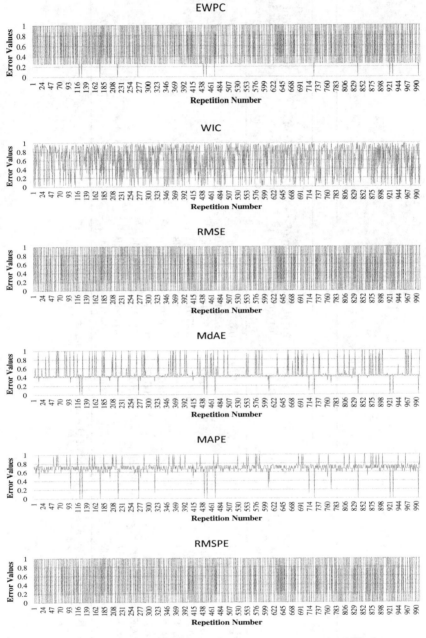

Fig. 4.   Graphs of error values of the performance criteria for MG dataset.

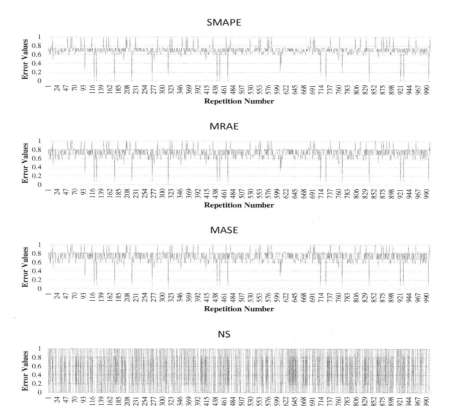

Fig. 4. (*Continued*)

the best forecasting performance with the proximity of 80.72971%. It is seen that 1-1-1 architecture are predominantly chosen by many performance criteria for SAR1. It is seen that the architectures with the smallest errors selected among the 1000 repetitions made vary widely in terms of performance. The WIC criterion has the highest volatility, and the SMAPE criterion has the lowest volatility.

Figure 5 shows the line graphs of the minimum error values obtained for 1000 repetitions for all the performance criteria of SAR1 (error values are rescaled to 0–1 range for comparison). It is seen that all criteria except for MRAE have similar characteristics compared to the others. Excluding MRAE, along with better performance, the error values are often around one for EWPC, MAPE, RMSPE and SMAPE and 0.8 for WIC, RMSE, MdAE, MASE.

Table 9. SAR1 results.

	Seasonal Autoregressive (SAR1)				
	Average of Correlation Values Between Test Set and Estimates	Mean Absolute Errors Between Test Set and Estimates	Average of One-Step-Ahead Forecasts, Actual Value = −0.6042	Percentage Proximity to One-Step-Ahead Real Value	Most Selected Architecture and Number of Selection
EWPC	0.9971	0.4746	−0.7653	73.34%	1-1-1/698
WIC	**0.9997**	0.4853	−0.7319	78.86%	1-1-1/1000
RMSE	0.9894	0.4686	−0.7270	79.68%	3-1-1/680
MdAE	0.8647	0.8994	−0.9635	40.53%	6-1-1/749
MAPE	0.9994	0.4842	−0.7333	78.64%	1-1-1/946
RMSPE	0.9996	0.4851	−0.7319	78.86%	1-1-1/960
SMAPE	0.9817	0.6282	−0.7429	77.04%	1-1-1/954
MRAE	0.8846	0.6044	−0.7206	**80.73%**	2-3-1/89
MASE	0.9670	**0.4573**	−0.9072	49.86%	6-1-1/652
NS	0.9894	0.4686	−0.7270	79.68%	3-1-1/680

Table 9. (*Continued*)

	Minimum Error Value	Architecture of the Minimum Error Value	Maximum Error Value	Mean and Median of Error Values Set to 0–1 Interval		Standard Deviation of Error Values Set to 0–1 Interval	Coefficient of Variation of Error Values Set to 0–1 Interval
				Seasonal Autoregressive (SAR1)			
EWPC	0.3502	7-3-1	0.4294	0.9208	1.0000	0.1381	15.0011
WIC	0.0480	1-1-1	0.0484	0.7050	0.8066	0.2331	33.0663
RMSE	0.4945	7-3-1	0.5990	0.8024	0.8748	0.1233	15.3637
MdAE	0.2033	5-2-1	0.3226	0.7353	0.7996	0.1560	21.2196
MAPE	0.2750	2-2-1	0.3031	0.9858	1.0000	0.1026	10.4037
RMSPE	0.3051	2-2-1	0.3056	0.8955	0.9999	0.2321	25.9157
SMAPE	0.3136	2-3-1	0.3569	0.9817	0.9993	0.0927	9.4462
MRAE	0.5684	6-4-1	1.0560	0.6716	0.6910	0.1717	25.5720
MASE	0.2809	7-3-1	0.3492	0.7928	0.8360	0.1157	14.5970
NS	0.1027	7-3-1	0.1507	0.7887	0.8644	0.1277	16.1901

Note: Bold entries are min-max values of the respective column.

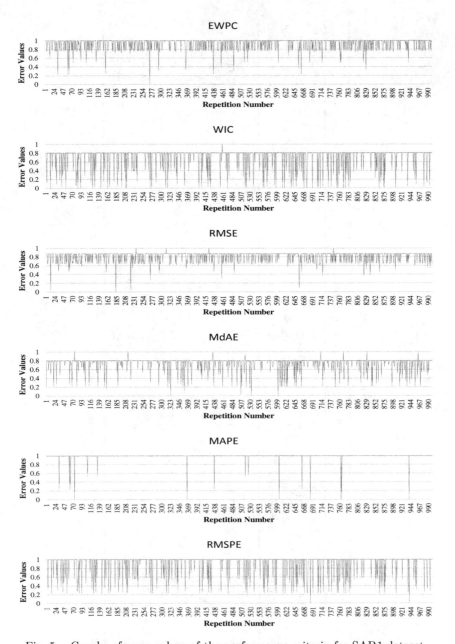

Fig. 5.   Graphs of error values of the performance criteria for SAR1 dataset.

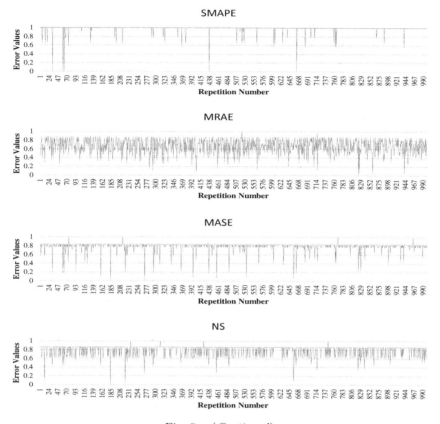

Fig. 5. (*Continued*)

The results of sinus time series are given in Table 10.

From Table 10, due to the complexity of the data for sinus, it is observed that it has very poor error performance with high mean absolute error and very low concordance performance with low correlation. However, it can be seen that RMSE and NS have the highest correlation and MASE has the lowest mean absolute error. On the other hand, it can be concluded that the forecasting performance is clearly poor for this data, as the one-step ahead forecasting value is far from the actual value of 0.0057. sinus reveals its complex structure in the selection of architecture, and it also causes the selection of very different architectures among all performance criteria.

Table 10.   Sinus results.

	Sinus				
	Average of Correlation Values Between Test Set and Estimates	Mean Absolute Errors Between Test Set and Estimates	Average of One-Step-Ahead Forecasts, Actual Value = 0.0057	Percentage Proximity to One-Step-Ahead Real Value	Most Selected Architecture and Number of Selection
EWPC	0.4661	0.2923	0.2952	−4979.1%	1-1-1/152
WIC	0.4723	0.3045	0.2674	−4490.8%	1-2-1/320
RMSE	0.5201	0.2955	0.3077	−5197.7%	1-7-1/106
MdAE	0.3155	0.2992	0.3022	−5102.2%	1-9-1/295
MAPE	0.3497	0.2948	0.2937	−4952.4%	1-5-1/221
RMSPE	0.3382	0.3100	0.2766	−4652.6%	1-2-1/420
SMAPE	0.3648	0.3187	0.2935	−4948.6%	9-2-1/89
MRAE	0.2215	0.4196	0.2238	−3726.9%	5-3-1/92
MASE	0.4257	**0.2813**	0.3085	−5212.3%	1-7-1/275
NS	0.5201	0.2955	0.3103	−5243.5%	1-7-1/106

Table 10.   (*Continued*)

Sinus

	Minimum Error Value	Architecture of the Minimum Error Value	Maximum Error Value	Mean and Median of Error Values Set to 0–1 Interval		Standard Deviation of Error Values Set to 0–1 Interval	Coefficient of Variation of Error Values Set to 0–1 Interval
EWPC	0.4907	1-5-1	0.6955	0.5651	0.5603	0.1529	27.0488
WIC	-0.0604	2-1-1	0.1911	0.3476	0.3535	0.1905	54.7945
RMSE	0.3325	12-3-1	0.4184	0.7076	0.7144	0.1595	22.5344
MdAE	0.1038	3-9-1	0.1864	0.5079	0.5184	0.1370	26.9661
MAPE	0.5647	1-5-1	0.9027	0.4984	0.5266	0.0894	17.9265
RMSPE	0.7199	1-4-1	1.1142	0.2662	0.2388	0.0867	32.5494
SMAPE	0.7455	6-6-1	0.9786	0.6954	0.6909	0.1491	21.4383
MRAE	0.3789	12-5-1	1.0213	0.5983	0.6083	0.1859	31.0731
MASE	0.4698	3-6-1	0.6003	0.5038	0.4980	0.1204	23.8925
NS	0.6220	12-3-1	0.9852	0.6869	0.6910	0.1647	23.9801

Note: Bold entries are min-max values of the respective column.

Thus, it can be interpreted that it is important to show how different the performance criteria handling of error can be.

The minimum error values, the architectures of the smallest error values and the coefficients of variation for the errors are given for each of the 1000 repetitions of all performance criteria for sinus. Sinus reveals its complex structure in the selection of the architecture with the smallest error in 1000 repetitions, and it is seen that it causes the selection of very different architectures among the criteria. The WIC criterion has the highest volatility, and the MAPE criterion has the lowest volatility.

Figure 6 shows the line graphs of the minimum error values obtained for 1000 repetitions for all the performance criteria of sinus (error values are rescaled to 0–1 range for comparison). It is thought that due to its complex structure, all the criteria have poor performance and all of them have similar characteristics in the graph of 1000 repetition errors. Although all the criteria show similar behavior, there may be slight differences between the MAPE, RMSPE and MASE criteria and the others among the error frequency ranges. It can be concluded that the criteria of MAPE, RMSPE and MASE have a narrower frequency of error, while other criteria tend to make a wider range of architectural choices.

The results of exponential time series are given in Table 11.

From Table 11, for the exponential dataset it can be said that, except for RMSPE, all the criteria have similar concordance performance. MASE has the best forecasting performance with the proximity of 97.79966%. Due to the complex structure of the exponential dataset, it can be said that the selection of architectures differs from case to case.

The minimum error values, architectures of the smallest error values and coefficients of variation for the errors are given for each of the 1000 repetitions of all performance criteria for exponential. It is seen that the architectures with the smallest error values selected among the 1000 repetitions made vary widely in terms of performance of each criterion. The WIC criterion appears to have a very high volatility.

Figure 7 shows the line graphs of the minimum error values obtained for 1000 repetitions for all the performance criteria of exponential (error values are rescaled to 0–1 range for comparison). It can be observed that all the criteria except the WIC criteria have similar

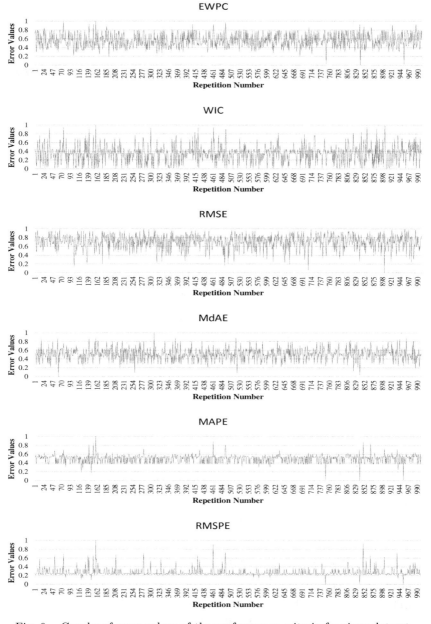

Fig. 6.   Graphs of error values of the performance criteria for sinus dataset.

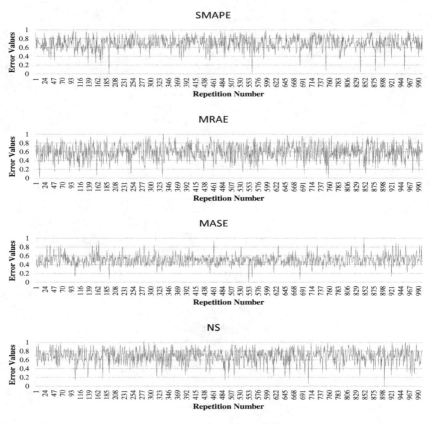

Fig. 6.   (*Continued*)

characteristics and also from the graphs that the WIC criterion is highly performing in architectural selection. Excluding MRAE, all criteria have similar performance but, in more detail, the error values are often around 1 for RMSE, MdAE, MASE and NS and 0.8 for EWPC, RMSPE and SMAPE.

3.2.   *Applications with real-world data*

The real-world datasets are exchange rates for USD/EUR, USD/GBP and USD/JPY. In addition, three well-known time series data are used. These are Yellowstone Park geyser explosions on the basis of minutes (Härdle, 1991), the annual number of Canadian Lynxes caught between the years 1821 and 1934 (Bulmer, 1974)

Table 11. Exponential results.

	Exponential			
	Average of Correlation Values Between Test Set and Estimates	Mean Absolute Errors Between Test Set and Estimates	Average of One-Step-Ahead Forecasts, Actual Value = −1.0967	Percentage Proximity to One-Step-Ahead Real Value	Most Selected Architecture and Number of Selection
EWPC	0.8922	0.3213	−1.0867	99.0887%	1-1-1/290
WIC	0.8928	0.3328	−1.0729	97.8282%	1-1-1/998
RMSE	0.9014	0.3164	−1.0988	99.8112%	3-1-1/307
MdAE	0.8589	0.4213	−1.0590	96.5595%	3-9-1/221
MAPE	0.8514	0.3642	−1.0449	95.2799%	1-2-1/200
RMSPE	0.8210	0.4051	−1.0516	95.8903%	1-2-1/273
SMAPE	0.8953	0.3096	−1.0721	97.7524%	5-1-1/431
MRAE	0.8608	0.3767	−1.1208	97.7997%	4-3-1/123
MASE	0.8997	**0.3054**	−1.0752	98.0373%	5-1-1/310
NS	0.9014	0.3164	**−1.0952**	99.8641%	3-1-1/307

(*Continued*)

Table 11. (*Continued*)

	Minimum Error Value	Architecture of the Minimum Error Value	Maximum Error Value	Exponential Mean and Median of Error Values Set to 0-1 Interval		Standard Deviation of Error Values Set to 0-1 Interval	Coefficient of Variation of Error Values Set to 0-1 Interval
EWPC	0.3162	2-4-1	0.4817	0.7374	0.8013	0.1888	25.5963
WIC	−0.1058	1-1-1	−0.0502	0.0030	0.0000	0.0547	1816.33
RMSE	0.2973	8-4-1	0.3907	0.9190	0.9500	0.1157	12.5919
MdAE	0.1017	4-4-1	0.2503	0.6308	0.6738	0.2347	37.2105
MAPE	0.3406	10-3-1	0.7140	0.7402	0.7846	0.1754	23.6887
RMSPE	0.4805	4-3-1	1.2351	0.5565	0.5620	0.1625	29.2085
SMAPE	0.3153	2-4-1	0.4269	0.7873	0.8742	0.1671	21.2176
MRAE	0.3687	5-6-1	0.7527	0.5083	0.5224	0.1860	36.5932
MASE	0.3337	8-4-1	0.4918	0.8154	0.8655	0.1540	18.8822
NS	0.1189	8-4-1	0.2053	0.9107	0.9436	0.1243	13.6441

Note: Bold entries are min values of the second column and the closest value to the actual value of the third column.

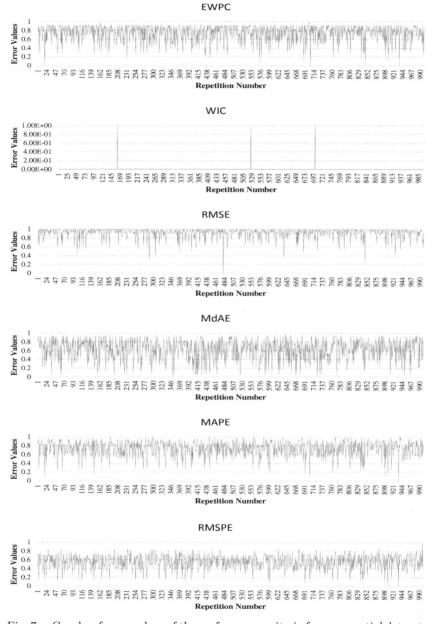

Fig. 7.   Graphs of error values of the performance criteria for exponential dataset.

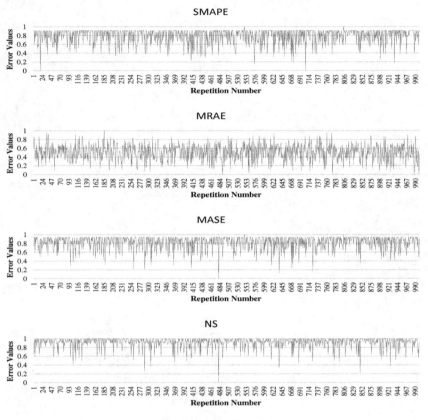

Fig. 7.   (*Continued*)

and the monthly number of airplane passengers obtained between 1949 and 1960 (Box and Jenkins, 1976).

The exchange rates data have been obtained from the official website (www.tcmb.gov.tr) of the Central Bank of the Republic of Turkey. They are weekly and cover totally 247 observations between March 23, 2012 and December 9, 2016. Weekly observation values are the average of daily parity values for each working day of the week.

The numbers of observations for the data of Canadian Lynxes, Yellowstone Park geyser explosions and aircraft passengers are 114, 272 and 144, respectively.

Fig. 8.   Graphs of USD/EURO and USD/GBP time series.

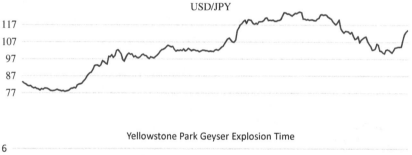

Fig. 9.   Graphs of USD/JPY and Yellowstone Park geyser explosions time series.

The scatter plots of the time series analyzed in this chapter are given in Figs. 8–10. From the figures, it can be seen that these series data have complex structures and different characteristics.

The results of USD/JPY time series are given in Table 12.

From Table 12, the other criteria except for MdAE and MRAE for the USD/JPY exchange rate time series have high concordance

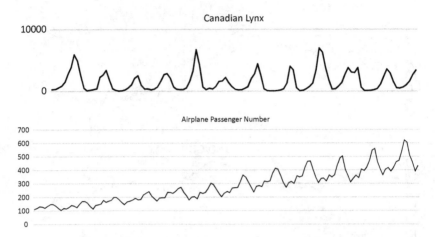

Fig. 10.   Graphs of Canadian Lynxes and airplane passengers number time series.

performance with very high average correlation. MASE has the best performance with the lowest average absolute error, and MRAE has the best forecasting performance with the proximity of 99.55440%. It is seen that 1-2-1 architecture is predominantly chosen by many performance criteria for USD/JPY.

The coefficients of variation for the errors of the 1000 networks are computed for each performance criterion. Apart from MdAE and MRAE, 1-2-1 architecture has the best performance after over 1000 repetitions. MdAE and MRAE have the highest and the lowest coefficients of variation, respectively.

Figure 11 shows the line graphs of the minimum error values obtained for 1000 repetitions for all the performance criteria of USD/JPY (error values are rescaled to 0–1 range for comparison). It is seen that MdAE and MRAE have different characteristics compared to the others. It is seen that MdAE has the highest variability and MRAE has a very low variability compared to the other criteria. EWPC, WIC, RMSE, MAPE, RMSPE, SMAPE, MASE and NS have very similar characteristics. The error values are often around 0.8 for WIC, 0.6 for RMSE, RMSPE and NS and 0.4 for MAPE, SMAPE and MASE. It is remarkable to point out that WIC cannot reach lower errors than the others.

The results of USD/EURO time series are given in Table 13.

From Table 13, the WIC criterion has the highest concordance performance with very high average correlation for the USD/EURO

Table 12.   USD/JPY results.

	USD/JPY			
	Average of Correlation Values Between Test Set and Estimates	Mean Absolute Errors Between Test Set and Estimates	Average of One-Step-Ahead Forecasts, Actual Value = 113.92	Percentage Proximity to One-Step-Ahead Real Value	Most Selected Architecture and Number of Selection
EWPC	1.0000	0.2120	113.2625	99.42%	1-2-1/712
WIC	1.0000	0.2143	113.2376	99.40%	1-2-1/549
RMSE	1.0000	0.2123	113.2490	99.41%	1-2-1/720
MdAE	0.9942	0.3565	**113.2441**	**99.41%**	1-8-1/291
MAPE	1.0000	0.2121	113.2663	99.43%	1-2-1/701
RMSPE	1.0000	0.2120	113.2517	99.41%	1-2-1/714
SMAPE	1.0000	0.2121	113.2663	99.43%	1-2-1/701
MRAE	0.9955	0.3444	113.4124	99.55%	3-1-1/824
MASE	1.0000	0.2120	113.2533	99.41%	1-2-1/716
NS	1.0000	0.2123	113.2490	99.41%	1-2-1/720

(Continued)

Table 12. (*Continued*)

				USD/JPY		
	Minimum Error Value	Architecture of the Minimum Error Value	Maximum Error Value	Mean and Median of Error Values Set to 0–1 Interval	Standard Deviation of Error Values Set to 0–1 Interval	Coefficient of Variation of Error Values Set to 0–1 Interval
EWPC	0.0756	1-2-1	0.1304	0.6545 / 0.4358	0.2673	40.8330
WIC	−0.6316	1-2-1	−0.4665	0.8832 / 0.7953	0.1138	12.8823
RMSE	0.1611	1-2-1	0.2509	0.7690 / 0.6272	0.1789	23.2631
MdAE	0.0203	1-8-1	0.2134	0.2049 / 0.1478	0.1752	85.4737
MAPE	0.0013	1-2-1	0.0023	0.6582 / 0.4410	0.2663	40.4651
RMSPE	0.0015	1-2-1	0.0024	0.7240 / 0.5518	0.2120	29.2822
SMAPE	0.0013	1-2-1	0.0023	0.6581 / 0.4409	0.2664	40.4728
MRAE	0.1440	2-2-1	0.2567	0.9741 / 1.0000	0.0771	7.9121
MASE	0.1054	1-2-1	0.1863	0.6777 / 0.4698	0.2553	37.6657
NS	0.0019	1-2-1	0.0045	0.7372 / 0.5762	0.2008	27.2360

Note: Bold entries are min-max values of the respective column.

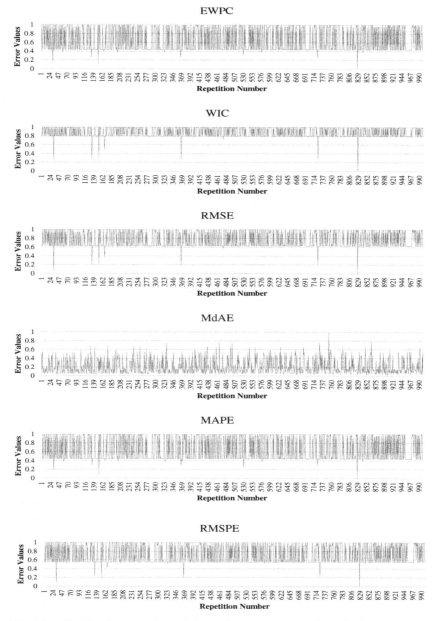

Fig. 11.   Graphs of error values of the performance criteria for USD/JPY dataset.

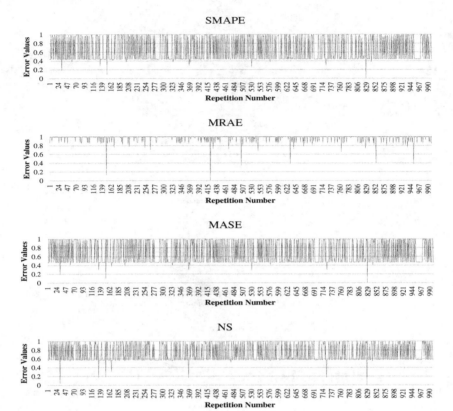

Fig. 11. (*Continued*)

exchange rate time series. However, it can be said that the WIC criterion, along with MdAE, has a higher mean absolute error value for 1000 repetitions than other criteria for USD/EURO and therefore they have lower performance. SMAPE has the best forecasting performance with the proximity of 99.87279%. It is seen that 12-1-1 architecture is predominantly chosen by many performance criteria for USD/EURO. The coefficients of variation for the errors of the 1000 networks are computed for each performance criterion. Apart from WIC and MdAE, 6-2-1 and 6-3-1 architectures have the best performance over 1000 repetitions. WIC and MASE have the highest and the lowest coefficients of variation, respectively.

Figure 12 shows the line graphs of the minimum error values obtained for 1000 repetitions for all the performance criteria of

Table 13. USD/EURO results.

	Average of Correlation Values Between Test Set and Estimates	Mean Absolute Errors Between Test Set and Estimates	Average of One-Step-Ahead Forecasts, Actual Value = 1.071	Percentage Proximity to One-Step-Ahead Real Value	Most Selected Architecture and Number of Selection
EWPC	0.9827	0.0032	1.0687	99.78%	12-1-1/505
WIC	**0.9934**	0.0037	1.0727	99.84%	2-1-1/1000
RMSE	0.9821	0.0032	1.0684	99.76%	12-1-1/581
MdAE	0.9192	0.0167	0.9747	91.01%	1-8-1/158
MAPE	0.9828	0.0031	1.0696	99.87%	12-1-1/390
RMSPE	0.9821	0.0032	1.0684	99.76%	12-1-1/583
SMAPE	0.9828	0.0031	1.0696	99.87%	12-1-1/399
MRAE	0.9826	0.0031	1.0696	**99.86%**	12-1-1/347
MASE	0.9826	**0.0031**	1.0696	99.86%	12-1-1/344
NS	0.9821	0.0032	1.0684	99.76%	12-1-1/581

(*Continued*)

Table 13.   (*Continued*)

	Minimum Error Value	Architecture of the Minimum Error Value	Maximum Error Value	USD/EURO Mean and Median of Error Values Set to 0–1 Interval		Standard Deviation of Error Values Set to 0–1 Interval	Coefficient of Variation of Error Values Set to 0–1 Interval
EWPC	0.0377	6-3-1	0.0531	0.9988	0.8102	0.2802	34.5878
WIC	−2.1124	2-1-1	−2.1122	0.4868	0.4669	0.0773	16.5622
RMSE	0.0029	6-3-1	0.0041	0.9991	0.8466	0.2383	28.1495
MdAE	0.0012	2-3-1	0.0029	0.7019	0.6701	0.1415	21.1109
MAPE	0.0022	6-2-1	0.0030	0.9569	0.7766	0.3068	39.5084
RMSPE	0.0026	6-3-1	0.0037	0.9991	0.8558	0.2283	26.6755
SMAPE	0.0022	6-2-1	0.0030	0.9555	0.7756	0.3069	39.5706
MRAE	0.0024	6-2-1	0.0034	0.9455	0.7762	0.3069	39.5410
MASE	0.2919	6-2-1	0.4126	0.9413	0.7725	0.3062	39.6336
NS	0.0230	6-3-1	0.0460	0.9989	0.8340	0.2518	30.1858

Note: Bold entries are min-max values of the respective column.

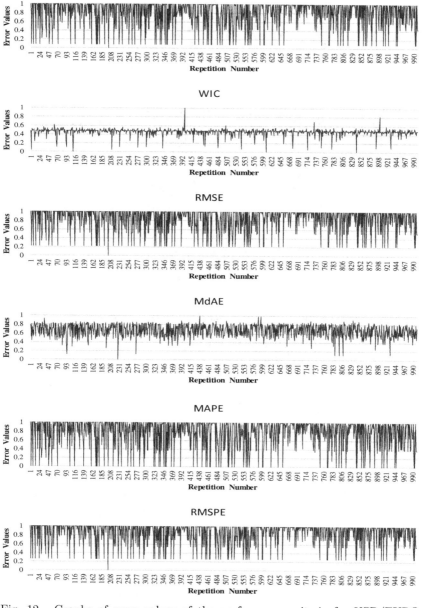

Fig. 12.   Graphs of error values of the performance criteria for USD/EURO dataset.

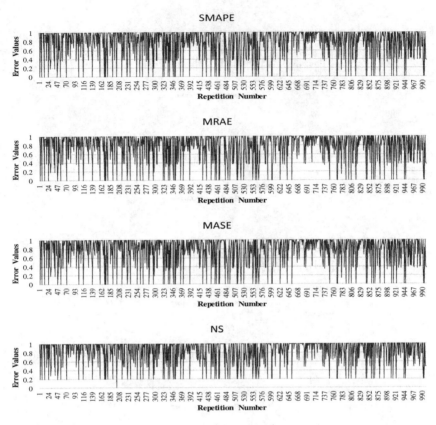

Fig. 12.   (*Continued*)

USD/EURO (error values are rescaled to 0–1 range for comparison). It is seen that WIC and MdAE have different characteristics compared to the others. Although WIC and MdAE criteria have lower coefficients of variation than other criteria, it can be seen that the frequency of the architectures with low error values is less than other criteria. It is seen that the criteria of RMSE, RMSPE and NS have very similar characteristics with each other and the frequency of architectures with low error values has decreased considerably after 0.2, and they can even make only one selection of architecture with the lowest error value for this study. On the other hand, although the coefficients of variation of MAPE, SMAPE, MRAE and MASE are higher than the other criteria, it is seen that the frequency of reaching the architectures with low error values is quite high. EWPC, which is a weighted criterion, has a higher coefficient of variation compared to the criteria of RMSE, RMSPE and NS, but it is observed that the frequency of architectures with low error values is around 0.1. Similarly, EWPC has a lower coefficient of variation than MAPE, SMAPE, MRAE and MASE, but the frequency of the architectures with low error values is lower than that of MAPE, SMAPE, MRAE and MASE.

The results of USD/GBP time series are given in Table 14.

From Table 14, the WIC criterion has the highest concordance performance with very high average correlation for the USD/GBP exchange rate time series. The MdAE criterion has a higher mean absolute error value of 1000 repetitions than other criteria for USD/GBP and therefore it has lower performance. RMSPE has the best forecasting performance with the proximity of 99.67702%. It is seen that 1-1-1 architecture is predominantly chosen by many performance criteria for USD/GBP.

The coefficients of variation for the errors of the 1000 networks are computed for each performance criterion. Apart from WIC and MdAE, 3-3-1 and 1-8-1 architectures have the best performance over 1000 repetitions. WIC and MdAE have higher coefficients of variation among other criteria.

Figure 13 shows the line graphs of the minimum error values obtained for 1000 repetitions for all the performance criteria of USD/GBP (error values are rescaled to 0–1 range for comparison). It is seen that WIC and MdAE have different characteristics compared to the others. Although the WIC and MdAE criteria have higher coefficients of variation than other criteria, it can be seen that the frequency of the architectures with low error values is more than other criteria. It is seen that the EWPC, RMSE, RMSPE and NS criteria have characteristics very similar to each other and although they have small coefficients of variation, there are very few architectures with low error values in 1000 repetitions. On the other hand, the fact that MAPE, SMAPE, MRAE and MASE have very small numbers of architectures with low error values explains that the coefficients of variation are much lower than those of the other criteria.

The results of the Yellowstone Park geyser explosion time series are given in Table 15.

As seen in Table 15, the Yellowstone Park geyser explosion dataset is very complex and shows poor performance by all criteria. Among

Table 14. USD/GBP results.

	Average of Correlation Values Between Test Set and Estimates	Mean Absolute Errors Between Test Set and Estimates	USD/GBP Average of One-Step-Ahead Forecasts, Actual Value = 1.44132	Percentage Proximity to One-Step-Ahead Real Value	Most Selected Architecture and Number of Selection
EWPC	0.9986	0.0026	1.4463	99.7%	1-1-1/981
WIC	0.9987	0.0026	1.4464	99.7%	1-1-1/1000
RMSE	0.9985	0.0026	1.4461	99.7%	1-1-1/960
MdAE	0.4490	0.0089	1.5715	91.0%	1-4-1/479
MAPE	0.9987	0.0026	1.4463	99.7%	1-1-1/999
RMSPE	0.9984	0.0026	1.4460	99.7%	1-1-1/946
SMAPE	0.9987	0.0026	1.4463	99.7%	1-1-1/999
MRAE	0.9987	0.0026	1.4463	99.7%	1-1-1/998
MASE	0.9986	**0.0026**	1.4463	99.7%	1-1-1/997
NS	0.9985	0.0026	1.4461	99.7%	1-1-1/960

Table 14. (*Continued*)

	Minimum Error Value	Architecture of the Minimum Error Value	Maximum Error Value	USD/GBP Mean and Median of Error Values Set to 0–1 Interval	Standard Deviation of Error Values Set to 0–1 Interval	Coefficient of Variation of Error Values Set to 0–1 Interval
EWPC	0.0362	3-3-1	0.0372	0.9807 0.9996	0.1365	13.9235
WIC	−2.0707	1-1-1	−2.0706	0.1163 0.0003	0.2602	223.7571
RMSE	0.0044	3-3-1	0.0050	0.9750 0.9999	0.1481	15.1941
MdAE	0.0001	1-4-1	0.0008	0.4065 0.5603	0.2474	60.8581
MAPE	0.0017	1-8-1	0.0017	0.9989 0.9999	0.0316	3.1654
RMSPE	0.0029	3-4-1	0.0035	0.9703 0.9999	0.1484	15.2974
SMAPE	0.0017	1-8-1	0.0017	0.9989 0.9999	0.0316	3.1654
MRAE	0.0025	1-8-1	0.0026	0.9988 0.9999	0.0317	3.1774
MASE	0.2118	1-8-1	0.2213	0.9986 0.9999	0.0321	3.2184
NS	0.0170	3-3-1	0.0220	0.9748 0.9999	0.1486	15.2458

Note: Bold entries are min-max values of the respective column.

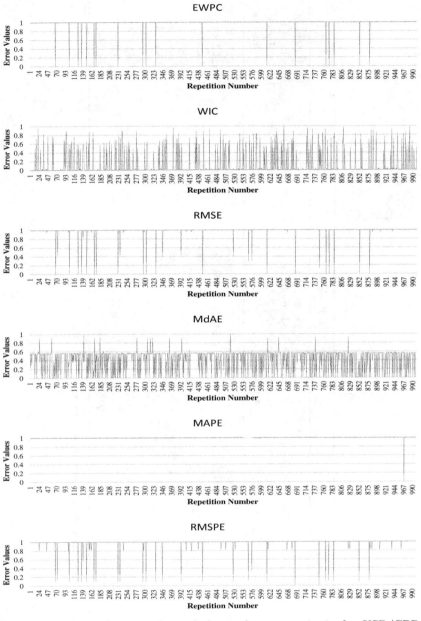

Fig. 13.   Graphs of error values of the performance criteria for USD/GBP dataset.

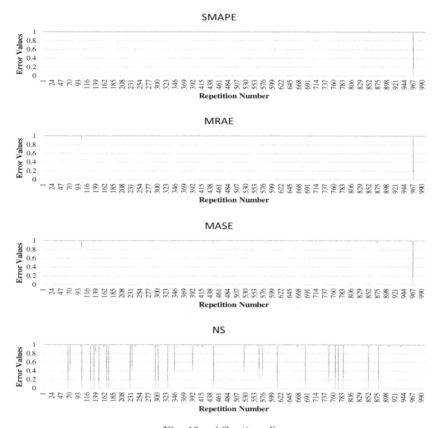

Fig. 13. (*Continued*)

the negative mean correlations, the WIC criterion has the worst value. MASE has the lowest average absolute error and SMAPE appears to be the one with the most average mean forecast at 81.78774%.

The minimum error values, architectures of the smallest error values and coefficients of variation for the errors are given for each of the 1000 repetitions of all the performance criteria for Yellowstone Park geyser explosions. It is seen that the 4-1-1 architecture is predominantly chosen by many performance criteria for Yellowstone Park geyser explosions.

Figure 14 shows the line graphs of the minimum error values obtained for 1000 repetitions for all the performance criteria of

Table 15.   Yellowstone Park geyser explosion results.

	Yellowstone Park Geyser Explosion				
	Average of Correlation Values Between Test Set and Estimates	Mean Absolute Errors Between Test Set and Estimates	Average of One-Step-Ahead Forecasts, Actual Value = 4.467	Percentage Proximity to One-Step-Ahead Real Value	Most Selected Architecture and Number of Selection
EWPC	−0.3257	1.2847	3.5415	79.28%	6-8-1/38
WIC	−0.8017	1.5188	3.0515	68.31%	1-1-1/586
RMSE	−0.4108	1.3043	3.4736	77.76%	6-5-1/35
MdAE	−0.2272	1.4018	**3.4881**	**78.09%**	6-10-1/50
MAPE	−0.2336	1.2997	3.4793	77.89%	6-8-1/39
RMSPE	−0.3123	1.3380	3.2698	73.20%	4-10-1/35
SMAPE	−0.2959	1.2887	3.6535	81.79%	6-8-1/52
MRAE	−0.4159	1.3572	3.1568	70.67%	4-5-1/27
MASE	−0.2742	1.2781	3.6164	80.96%	6-8-1/51
NS	−0.4108	1.3043	3.4736	77.76%	6-5-1/35

Table 15. *(Continued)*

Yellowstone Park Geyser Explosion

	Minimum Error Value	Architecture of the Minimum Error Value	Maximum Error Value	Mean and Median of Error Values Set to 0–1 Interval		Standard Deviation of Error Values Set to 0–1 Interval	Coefficient of Variation of Error Values Set to 0–1 Interval
EWPC	0.6331	4-1-1	1.1051	0.6888	0.7229	0.1785	25.9102
WIC	0.4728	4-1-1	0.9833	0.8812	1.0000	0.2038	23.1319
RMSE	1.0553	6-1-1	1.6906	0.6924	0.7353	0.1806	26.0796
MdAE	0.6006	10-12-1	1.3499	0.6096	0.6296	0.1492	24.4679
MAPE	0.3277	4-1-1	0.5176	0.6465	0.6653	0.1603	24.7963
RMSPE	0.4181	4-1-1	0.6605	0.6763	0.7043	0.1613	23.8468
SMAPE	0.2876	4-1-1	0.4332	0.6638	0.6913	0.1722	25.9399
MRAE	0.8027	7-1-1	1.2352	0.6589	0.6941	0.1861	28.2378
MASE	0.5697	4-1-1	0.8926	0.6597	0.6884	0.1703	25.8098
NS	0.9378	4-1-1	2.4067	0.6507	0.6903	0.1872	28.7697

Note: Bold entries are min-max values of the respective column.

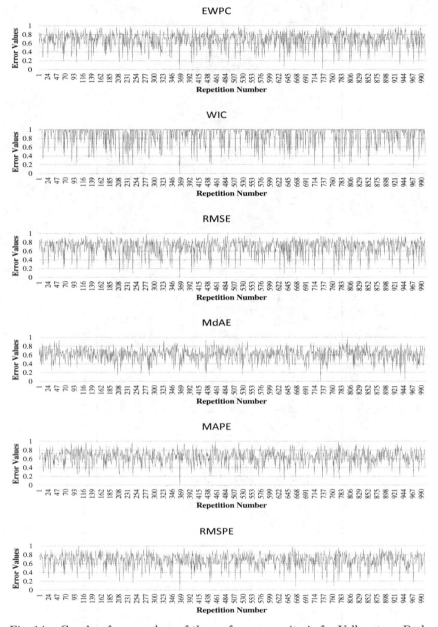

Fig. 14.   Graphs of error values of the performance criteria for Yellowstone Park geyser explosion dataset.

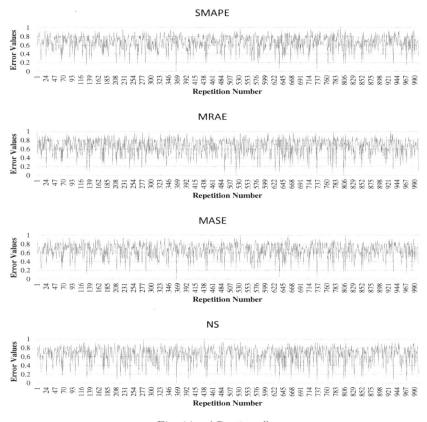

Fig. 14. (*Continued*)

Yellowstone Park geyser explosions (error values are rescaled to 0–1 range for comparison). It can be observed that all the criteria except WIC have similar characteristics and also from the graphs that WIC has a lower performance than other criteria in architecture selection.

The results of Canadian Lynx time series are given in Table 16. From Table 16, due to its uncomplicated structure, it is possible to say that all the criteria except for the MdAE criterion have high concordance and error performance. However, it is seen that the MdAE criterion shows the best forecasting performance with 96.63246%

Table 16.   Canadian Lynx results.

	Average of Correlation Values Between Test Set and Estimates	Mean Absolute Errors Between Test Set and Estimates	Canadian Lynx Average of One-Step-Ahead Forecasts, Actual Value = 3396	Percentage Proximity to One-Step-Ahead Real Value	Most Selected Architecture and Number of Selection
EWPC	0.9938	0.0163	3095.13	91.14%	1-1-1/1000
WIC	**0.9938**	0.0163	3095.13	91.14%	1-1-1/1000
RMSE	0.9938	0.0163	3095.13	91.14%	1-1-1/1000
MdAE	0.9247	0.2656	3281.64	96.63%	1-4-1/419
MAPE	0.9938	0.0163	3095.13	91.14%	1-1-1/1000
RMSPE	0.9938	0.0163	3095.13	91.14%	1-1-1/1000
SMAPE	0.9938	0.0163	3095.13	91.14%	1-1-1/1000
MRAE	0.9938	**0.0163**	3095.13	**91.14%**	1-1-1/1000
MASE	0.9938	0.0163	3095.13	91.14%	1-1-1/1000
NS	0.9938	0.0163	3095.13	91.14%	1-1-1/1000

Table 16. *(Continued)*

Canadian Lynx

	Minimum Error Value	Architecture of the Minimum Error Value	Maximum Error Value	Mean and Median of Error Values Set to 0–1 Interval		Standard Deviation of Error Values Set to 0–1 Interval	Coefficient of Variation of Error Values Set to 0–1 Interval
EWPC	0.0619	1-1-1	0.0619	0.6830	0.6609	0.2735	40.0493
WIC	−1.5239	1-1-1	−1.5239	0.6831	0.6611	0.2735	40.0423
RMSE	0.0176	1-1-1	0.0176	0.6831	0.6611	0.2735	40.0431
MdAE	0.0089	12-4-1	0.0189	0.7581	0.6706	0.1727	22.7751
MAPE	0.0281	1-1-1	0.0281	0.6837	0.6619	0.2735	40.0028
RMSPE	0.0302	1-1-1	0.0302	0.6836	0.6617	0.2735	40.0137
SMAPE	0.0278	1-1-1	0.0278	0.6837	0.6619	0.2735	40.0030
MRAE	0.0162	1-1-1	0.0162	0.6836	0.6617	0.2735	40.0125
MASE	0.3823	1-1-1	0.3823	0.6836	0.6617	0.2735	40.0141
NS	0.0433	1-1-1	0.0433	0.6831	0.6611	0.2735	40.0432

Note: Bold entries are min-max values of the respective column.

despite the poor concordance and error performance. It is seen that the same applies to the selection of architecture and 1-1-1 architecture is chosen by all other criteria except MdAE.

The minimum error values, architectures of the smallest error values and coefficients of variation for the errors are given for each of the 1000 repetitions of all the performance criteria for Canadian Lynx. It is seen that 1-1-1 architecture has been selected as the architecture with the smallest error by all criteria except the MdAE criterion. It can be concluded that there is little difference between the coefficients of variation calculated after bringing the errors to the 0–1 range, except for the MdAE criterion.

Figure 15 shows the line graphs of the minimum error values obtained for 1000 repetitions for all the performance criteria of Canadian Lynx (error values are rescaled to 0–1 range for comparison). It can be observed that all criteria except MdAE criteria have almost the same characteristics and also from the graphs that the MdAE criterion has lower performance in architecture selection compared to other criteria, that is, the frequency of reaching the architectures with small errors is less.

The results of aircraft passengers time series are given in Table 17.

From Table 17, due to its uncomplicated structure, it is possible to say that the high-performance architecture selections are made by all criteria for the aircraft passengers data, but it is quite surprising that the MdAE criterion with negative average correlation also has the best average forecasting and very high forecasting performance with 98.76651%. The highest mean correlation value is obtained by SMAPE, and the lowest mean absolute error value is provided by MASE criterion and similarly high performance is obtained by selecting the 12-1-1 architecture in high repetition by many criteria in architecture selection.

Figure 16 shows the line graphs of the minimum error values obtained for 1000 repetitions for all the performance criteria of aircraft passengers (error values are rescaled to 0–1 range for comparison). It is seen that the MdAE criterion has different characteristics than other criteria. EWPC, WIC, RMSE, MAPE, RMSPE,

EWPC

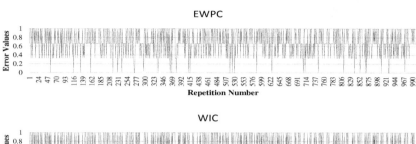

WIC

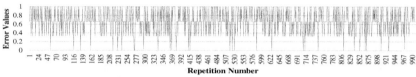

RMSE

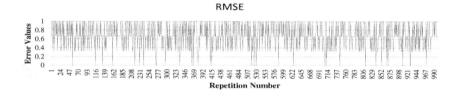

MdAE

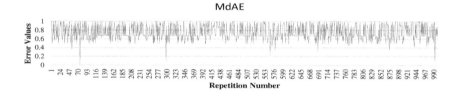

MAPE

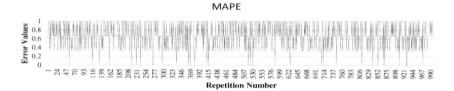

RMSPE

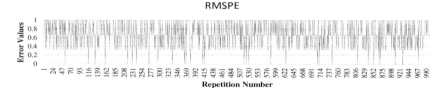

Fig. 15.   Graphs of error values of the performance criteria for Canadian Lynx dataset.

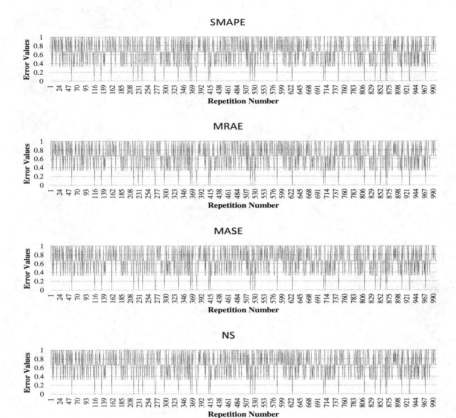

Fig. 15. (*Continued*)

SMAPE, MASE and NS criteria have very similar characteristics, but the frequency of the small error architectures reached 0.1 for WIC and 0.2 for EWPC, RMSE and NS criteria and 0.3 for RMSE, MAPE, SMAPE and MASE criteria and finally 0.4 for the MRAE criterion.

Table 17.   Aircraft passengers results.

	Aircraft Passengers				
	Average of Correlation Values Between Test Set and Estimates	Mean Absolute Errors Between Test Set and Estimates	Average of One-Step-Ahead Forecasts, Actual Value = 432	Percentage Proximity to One-Step-Ahead Real Value	Most Selected Architecture and Number of Selection
EWPC	0.9753	14.8426	421.0234	97.5%	12-1-1/940
WIC	0.9757	14.8562	421.0833	97.5%	12-1-1/971
RMSE	0.9753	14.8434	421.0386	97.5%	12-1-1/939
MdAE	−0.3569	13083.9658	426.6713	98.8%	1-3-1/324
MAPE	0.9757	14.8344	421.0919	97.5%	12-1-1/940
RMSPE	0.9757	14.8426	421.0563	97.5%	12-1-1/947
SMAPE	0.9758	14.8365	421.0916	97.5%	12-1-1/941
MRAE	0.9748	14.8910	420.8976	97.4%	12-1-1/926
MASE	0.9756	**14.8302**	421.0859	97.5%	12-1-1/937
NS	0.9753	14.8434	421.0386	97.5%	12-1-1/939

(*Continued*)

Table 17. (*Continued*)

Aircraft Passengers

	Minimum Error Value	Architecture of the Minimum Error Value	Maximum Error Value	Mean and Median of Error Values Set to 0–1 Interval		Standard Deviation of Error Values Set to 0–1 Interval	Coefficient of Variation of Error Values Set to 0–1 Interval
EWPC	12.3163	12-1-1	57.6410	0.1852	0.1728	0.1007	54.4004
WIC	4.6005	12-1-1	7.8673	0.1281	0.1114	0.1175	91.7046
RMSE	13.1417	12-1-1	31.0449	0.2474	0.2362	0.0949	38.3657
MdAE	0.4888	1-12-1	14.4135	0.5938	0.5819	0.2041	34.3722
MAPE	0.0248	12-3-1	0.0497	0.2748	0.2647	0.0945	34.3842
RMSPE	0.0298	12-3-1	0.0616	0.2285	0.2160	0.0999	43.7276
SMAPE	0.0249	12-3-1	0.0483	0.2888	0.2787	0.0951	32.9446
MRAE	0.3419	12-3-1	0.9195	0.3892	0.3856	0.0799	20.5393
MASE	0.2369	12-3-1	0.5126	0.2780	0.2683	0.0932	33.5448
NS	0.0336	12-3-1	0.1872	0.1756	0.1631	0.1011	57.5932

Note: Bold entries are min-max values of the respective column.

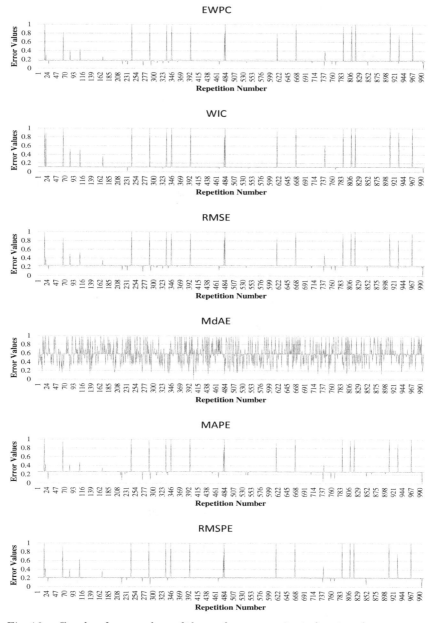

Fig. 16. Graphs of error values of the performance criteria for aircraft passengers dataset.

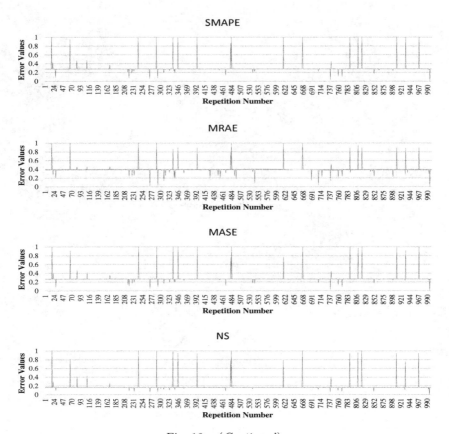

Fig. 16.　(*Continued*)

4.　Conclusion

The selection of the most appropriate network is basically the selection of the network with the smallest test set error in terms of error performance, but it can also mean the selection of the network with the best concordance performance or forecasting performance. While the basis of statistical modeling is defined as obtaining the model that gives the smallest error, ANN is based on the selection of the architecture with the smallest error among the obtained ones by trial-and-error method which are used frequently. Selecting the most appropriate architecture among the models created by ANN is a very important step toward reaching a satisfactory result. Since each criterion used for selection has different characteristics, ANN evaluates this process with different approaches. This is aimed to achieve rational results by using the weighted criteria which are formed by keeping the advantages and disadvantages of the various criteria together at an average level.

In this study, it is aimed to introduce the performance criteria available in the literature and to find out in which situations they perform better. In addition, EWPC, which is formed from the combination of various commonly used criteria, has been proposed as a weighted performance measure. Four different simulated datasets, three different foreign exchange rates time series, and three different well-known real-life datasets were applied to compare the performances of all criteria.

When the results for the weighted criteria are evaluated in terms of concordance, error, and forecast performance, the WIC criterion has the same average correlation value as the EWPC criterion in the USD/JPY, MG and Canadian Lynx datasets. For the Yellowstone Park geyser explosion dataset, WIC has a lower average correlation value than EWPC. However, it can be concluded that WIC has a higher mean correlation value for other datasets and therefore better concordance performance.

It was concluded that EWPC has a better performance due to lower average absolute error than WIC, except for the USD/GBP and Canadian Lynx datasets. The detailed information has been given individually, therefore this brief result is quite enough to use like this. However, as seen from the results of the detailed analysis, EWPC has 11% higher forecasting performance than WIC for the Yellowstone Park geyser explosion data. For the SAR1 data, WIC shows 5% better forecasting performance than EWPC. The coefficients of variation of the error values obtained over 1000 repetitions and adjusted to the 0–1 range show that WIC has a lower variability in half of the 10 datasets analyzed and EWPC has a lower variability for the remaining half of the 10 datasets. However, when the differences between the coefficients of variation are examined, it is noticed that WIC has 16 times higher variability than EWPC for USD/GBP data and it has 70 times higher variability than EWPC for the exponential data. From the graphs, it can be seen that WIC makes better architecture selection than EWPC. Finally, it can be concluded that the two criteria have different characteristics in terms of selecting architectures.

If it is necessary to make a general evaluation with the results obtained, it can be stated that the MdAE criterion has very low concordance and error performance, but along with MRAE and SMAPE, MdAE has quite a high forecasting performance. It can easily be seen that WIC is better than the other criteria in terms of concordance performance, and it is also easily seen that the MASE criterion has the best error performance for all datasets examined. When the coefficients of variation of the errors over 1000 repetitions are examined, it is seen that the WIC and MdAE criteria have very high variability for some datasets (which is smaller-better type of statistical indicator) compared to other criteria.

When the graphs of the errors are examined, it is seen that EWPC provides satisfactory results for all datasets in terms of both the selection of alternative architecture and frequency of the architectures with small error values. In terms of concordance, error and forecasting performances, it can be concluded that EWPC does not have the best or worst performance for any dataset and tends to perform close to the criterion which has the best performance for each data.

The use of weighted criteria, such as WIC and EWPC, can be very advantageous for selecting the appropriate network as they contain the characteristics of many criteria. In conclusion, it is recommended that the weighted criteria should be used by considering the characteristics of the data in order to be more effective at architecture selection in ANN.

References

Aladağ, C.H., Egrioglu, E. and Kadilar, C. (2010). Modeling brain wave data by using artificial neural. *Journal of Mathematics and Statistics*, **39**(1), 81–88.

Aladağ, C.H., Egrioglu, E., Gunay, S. and Basaran, M.A. (2010). Improving weighted information criterion by using optimization. *Journal of Computational and Applied Mathematics*, **233**(10), 2683–2687. https://doi.org/10.1016/j.cam.2009.11.016.

Bal, C. and Demir, S. (2017). Forecasting TRY/USD exchange rate with various artificial Neural Network Models. *TEM Journal*, **6**(1), 11–16. https://doi.org/10.18421/TEM61-02.

Bal, C., Demir, S. and Aladağ, C.H. (2016). A comparison of different model selection criteria for forecasting EURO/USD exchange rates by feed forward neural network. *International Journal of Computing, Communications & Instrumentation Engineering,* **3**(2), 1–5.

Box, G.E.P. and Jenkins, G.M. (1976). *Time Series Analysis: Forecasting and Control (Revised Edition),* (San Franciso, CA: Holden-Day).

Bulmer, M.G. (1974). A statistical analysis of the 10-year cycle in Canada. *Journal of Animal Ecology,* **43**(3), 701–718. https://doi.org/10.2307/3532.

Cybenko, G. (1989). Approximation by superpositions of a sigmoidal function. *Mathematics of Control, Signals and Systems,* **2**(4), 303–314. https://doi.org/10.1007/BF02551274.

Eğrioğlu, E., Aladağ, Ç.H. and Günay, S. (2008). A new model selection strategy in artificial neural networks. *Applied Mathematics and Computation,* **195**(2), 591–597. https://doi.org/10.1016/j.amc.2007.05.005.

Fushiki, T. (2011). Estimation of prediction error by using K-fold cross-validation. *Statistics and Computing,* **21**(2), 137–146. https://doi.org/10.1007/s11222-009-9153-8.

Glass, L. and Mackey, M. C. (1979). Pathological conditions resulting from instabilities in physiological control systems.* *Annals of the New York Academy of Sciences,* **316**(1), 214–235. https://doi.org/10.1111/j.1749-6632.1979.tb29471.x.

Härdle, W. (1991). *Smoothing Techniques: With Implementation in S,* (New York: Springer).

Hornik, K. (1991). Approximation capabilities of multilayer feedforward networks. *Neural Networks,* **4**(2), 251–257. https://doi.org/https://doi.org/10.1016/0893-6080(91)90009-T.

Hyndman, R.J. and Koehler, A.B. (2006). Another look at measures of forecast accuracy. *International Journal of Forecasting,* **22**(4), 679–688. https://doi.org/10.1016/j.ijforecast.2006.03.001.

Koehler, A. (2001). The asymmetry of the sAPE measure and other comments on the M3Competition. *International Journal of Forecasting,* **17**.

Krogh, A. and Vedelsby, J. (1995). Neural network ensembles, cross validation, and active learning. *Advances in Neural Information Processing Systems,* 231–238.

McCulloch, W.S., and Pitts, W. (1943). A logical calculus of the ideas immanent in nervous activity. *The Bulletin of Mathematical Biophysics,* **5**(4), 115–133. https://doi.org/10.1007/BF02478259.

Nash, J.E. and Sutcliffe, J.V. (1970). River flow forecasting through conceptual models Part I — A discussion of principles. *Journal of Hydrology,* **10**(3), 282–290. https://doi.org/https://doi.org/10.1016/0022-1694(70)90255-6.

Rissanen, J. (1980). Consistent order estimates of autoregressive processes by shortest description of data. *Analysis and Optimisation of Stochastic Systems,* (New York: Academic Press).

Vogl, T.P., Mangis, J.K., Rigler, A.K., Zink, W.T. and Alkon, D.L. (1988). Accelerating the convergence of the back-propagation method. *Biological Cybernetics,* **59**(4–5), 257–263. https://doi.org/10.1007/BF00332914.

Werbos, P.J. (1974). Beyond regression: New tools for prediction and analysis in the behavioral sciences. *Foundations, PhD Thesis.* https://doi.org/10.1.1.41.8085.

Zaremba, W., Sutskever, I. and Vinyals, O. (2014). *Recurrent Neural Network Regularization.* **2013**, 1–8.

Zhang, G., Eddy Patuwo, B. and Y. Hu, M. (1998). Forecasting with artificial neural networks. *International Journal of Forecasting,* **14**(1), 35–62. https://doi.org/10.1016/S0169-2070(97)00044-7.

Zur, R.M., Jiang, Y., Pesce, L.L. and Drukker, K. (2009). Noise injection for training artificial neural networks: A comparison with weight decay and early stopping. *Medical Physics,* **36**(10), 4810–4818. https://doi.org/10.1118/1.3213517.

Index

variance of the measurement error, 41
variance–covariance matrix, 40
vector error correction, 52, 170
vertical line, 83
Viennet's test problem, 182
visual tool, 83
volatile rise, 48
volatility, 145

W

Ward method, 223
wavering product, 74
weather conditions, 152
Weibull distribution, 106
weight coefficients, 164
weight columns, 176
weight matrix, 40
weighted criteria, 240, 291–292
weighted criterion, 273
weighted errors, 239
weighted information criterion, 240

weighted performance measure, 291
Weighted Superposition
    Attraction–Repulsion, 175–176
white noise, 58
white swan, 132
Whitney's theorem, 72
Wiener filter, 87
wind, 136
World Bank database, 11, 223
World Development Indicator,
    135
World Economic Forum, 131
World Resources Institute, 154

X

XOR problem, 191–192

Y

Yahoo Finance, 166
yearly country factor scores, 14
youth on the move, 3, 21